From Conflict to Collaboration

Building Peace in Nigeria's

Oil-Producing Communities

Published by
Adonis & Abbey Publishers Ltd
P.O. Box 43418
London
SE11 4XZ
http://www.adonis-abbey.com
First Edition, April 2005

British Library Cataloguing-in-Publication Data

A catalogue record for this book is available from the British Library

ISBN: 1-905068-05-0

Cover Design   Ifeanyi Adibe

Printed and bound in Great Britain by Lightning Source UK Ltd.

From Conflict to Collaboration

Building Peace in Nigeria's

Oil-Producing Communities

By

Austin Onuoha

Adonis & Abbey
Publishers Ltd

Other Books by Adonis & Abbey include:

Broken Dreams by Jideofor Adibe

Wooden Gongs and Drumbeats: African Folktales, Proverbs and Idioms by Dahi Chris Onuchukwu

Nigeria and the Politics of Unreason by Victor E. Dike

The Challenge of Authenticity: African Culture and Faith Commitment by Jacob Hevi

The Spirit Sets Free by Jacob Hevi

Internationalisation and Enterprise Development in Ghana edited by John Kuada

Journals from Adonis & Abbey Publishers Ltd

- **African Renaissance**

African Renaissance is a bi-monthly publication targeting policymakers, policy moulders, academics, corporate executives and stakeholders on Africa. It is a cross between an academic publication and any higher-end news features magazine. Put simply, it is where Report Meets Analysis.

- **African Journal of Political and Social Research, AJPSR**

AJPSR is a peer-reviewed, international academic publication aimed at fostering social science research on Africa. It is not affiliated to any institution, nor is it driven by any funding agency. It is expected to debut later this year (2005).

- **African Journal of Economic and Business Research, AJEBR**

AJEBR is an independent, peer-reviewed academic journal aimed at fostering economic and business research on Africa. It is expected to debut this year (2005).

This book is dedicated to Rev. Fr. Kevin O'Hara, a priest, a human being   and a gentleman.

I am grateful to Cordaid for providing the grant used in preparing this work for publication.

Table of Contents

Acknowledgement

In summer 2003, while I was in Nigeria updating the data for this book, my brother in-law Maduabuchi, 22, was murdered in cold blood. Winter 2004, while concluding research for this book my father died. One week after my return to the US to complete the writing, armed bandits attacked my wife and two young kids in my home. As I was concluding the writing of this book, my sister-in-law Eunice Ngaju died. To transcend all these, I wrote. This book in a sense is trauma-healing for me.

There are so many people who have helped me in pursuing and defining who I am. My brother Daniel Onuoha started it all. I thank you for introducing me to the written word. I am indebted to Rev. Fr. Kevin O'Hara, founder and director of the Human Rights Commission, Abakaliki, and the Centre for Social and Corporate Responsibility in Port Harcourt, my boss, my friend, a caring human being. Sr. Rita Schwarzenberger ("godmother No.1") formally introduced me to the field of conflict resolution. Traudl Ott is the first "red head" I met in my life and since then it has not been the same. She was my "third party intervener in my conflict with Misereor". Her intervention secured for me the grant for my graduate studies at the Conflict Transformation Program (CTP) Eastern Mennonite University (EMU), Harrisonburg, U.S.A. She is a committed friend and colleague. I am grateful to the good men, women and children that support Misereor's work with their money. I am a testimony of your sacrifice.

The Conflict Transformation Program (CTP) of the Eastern Mennonite University (EMU) was quite a challenge. My first impression of CTP, EMU was through the ever-present Janelle Myers-Benner, she answers every query with a smile. Don Foth gave me my first dinner in the US, I will never forget that. His wife Margaret Foth, a graceful woman, arranged my first speaking engagement with the Rotary Club in Harrisonburg and I spoke on the title of this book. At EMU Vernon Jantzi taught me self-confidence, Howard Zehr taught me humility in achievement, Ron Kraybil taught me integrity with simplicity, Ruth Zimmerman (she knows everything) taught me "quiet efficiency". Jan Jenner taught me to look out for what worked.

The "Queen of the Universe" the irrepressible Jayne Docherty ("godmother No. 2") supervised this work. Thank you for being very patient with an adult learner. To my very good friend and colleague Jebiwot Sumbeiywo (J.B.), she asked me to follow this through. I "stole" Kathy Orovwigho's file to update my research for this book. Thanks for not pressing charges.

To my professors at the Department of Conflict Analysis and Resolution (DCAR) at Nova Southeastern University, I will always remain grateful especially to Dr. Judith McKay (the Chair of the Department), Dr. Marie Olson-Lounsbery, Dr. Hamdesa Tuso, and Dr. Marcia Sweedler. And to my fellow students, Andrew Rusatsi, Jacques Koko, Annette Anderson-Engler, Hilda Dunkwu, Cyril, Moussa and Josiah I say thank you.

I am also grateful to my friends and colleagues at the Human Rights Commission and Center for Social and Corporate Responsibility, Dr. David Okwudili, Dr E. O. Emmanuel, Paulinus (my resident critic), Eva, Moses, Jude, Patrick, Phil, Uzo, Abba, Fred and others too numerous to mention. And to all the good people of the Niger Delta especially the people of Ogoni, Ogbodo,

Batan, Umuechem, Ndoki and Akwete I say never lose hope!

And to my mum Mrs. Esther Akuba Onuoha, without a mother like you, what will the world be like? And to my late father Gabriel Onuoha, thank you for "giving that which you never had". And to my late brother Caleb Onuoha, my sister Beatrice Onuoha-Odikaesieme, Emma, and Nathan thank you for being there.

And finally to my friend, sister, "mother" and wife Sister Ngo, without your love, support and care, this would not have been possible. To you, all my love. And to my son Iheabunike and my beautiful "rat" Ndudi thanks for keeping mummy busy in daddy's absence.

Austin Onuoha
Fort Lauderdale, Florida
August, 2004

Introduction

In January 1996, I was on my way to Warri[1] when suddenly a young man barely in his twenties emerged from the bush and waved us to a stop. Strapped to his bare chest was an assault rifle. We panicked but our driver reassured us that the boy was not an armed robber. He ordered us to disembark and we were herded into the nearby bush. At an open spot there stood another group of about seventeen other teenagers. They all had the same type of rifle. We were asked to identify ourselves. As we did, we were dispossessed of all our valuables.

When it got to my turn, I showed them the identity card from the media house where I had worked. The leader of the group examined it carefully and asked me to move to another side away from the rest. Nothing was removed from me. I was not even searched. Noticing my surprise, the leader of the gang, as if to answer my unasked question said, your newspaper supports our cause.

Then he addressed the whole of us in impeccable Queen's English. "We are not robbers; we are fighting for our rights. We are sorry for making you victims but we have no choice". After the search, they herded us back to our vehicle; we boarded and continued our journey. As we moved away from the scene, the driver explained to us that the boys are from the Niger Delta, and that they were agitating for their rights. He said that this happen always and that travellers paid to avoid harassment.

As we continued on the journey some questions kept nagging my mind: If these boys are caught they will be executed as armed robbers. They may never tell their story. Even if they did who will listen? What of the innocent people who have been robbed? Can they ever find a justification for what happened to them?

For the very first time in my life I realized the magnitude of the Niger Delta problem. The issue of the Niger Delta is not a new one at least to most Nigerians. But most people read about it, some heard of it, others wrote about it, but I have just experienced it. I am from the Niger Delta[2]. Oil was found in my village in the early nineties and what is today known as the Niger Delta except Edo, Ondo and Delta states used to be part of the former eastern region[3] of Nigeria. I have a stake in the Niger Delta. I have a stake in the survival of Nigeria. And as a human rights activist I believe in justice.

My main motivation for this research project is to offer a fresh alternative in the pursuit of peace in the Niger Delta. Second, to offer a blueprint for strategic intervention in the Niger Delta. Third, I expect that this work will be a clarion call for the people of ND to come together and speak with one voice on their plight. Four, I hope that this will also provide an impetus for the more discerning peace-builders to pay more careful and detailed attention to the ND. Because of the location of the ND most of the conflicts there receive very little analytical and methodological attention.

Since the events of September 11, 2001 and its aftermaths, there appears to be a major policy shift in the US centering on how to meet the shortfall in crude oil supplies at a competitive price. Many nations are looking up to Africa, especially the Gulf of Guinea to augment their supplies. In the Gulf of Guinea, Nigeria is thought to have more than 30 billion barrels of crude oil reserves. But because of conflict, violence and the rise of ethnic militias, crude oil supplies from Nigeria has been erratic. This book explores the causes, sources and dynamics of the conflicts between the oil-bearing communities and oil companies. This is against the backdrop of huge investments in conflict resolution initiatives in the area.

The main thesis of this book is that instead of "a conflict", there are "conflicts" in the Niger Delta region. And that these conflicts are hierarchical in structure and needs to be transformed as such. Utilizing a multi-modal approach of analysis, the author develops a cross-level model of emancipatory conflict transformation and peace-building termed Community Education and Institution-building (CEIB). The role of what the author calls "critical constituency" (CC) is also discussed within the framework of CEIB. This model mainstreams a combination of action research, appreciative inquiry and elicitive model of peacebuilding.

This book explores the conflicts between the oil companies and the host communities from a conflict transformation and peacebuilding perspective. Its uniqueness lies in the fact that it combines the theoretical postulations of prominent conflict theorists, with qualitative data and hands-on experience to do an in depth analysis of the conflicts there. It is not written from an economic, political, environmental or human rights perspective. The book looks at the conflicts in the Niger Delta from a social-interaction paradigm thereby divorcing it from history of the evolution of the Nigerian nation. The point of departure for this book is that it posits that the conflicts in the Niger Delta are embedded in the triangular relationship, which exists between the government, the oil companies and the host communities. It also makes the case for a more methodological framework of analysis that locates the conflicts within the social systems of the communities, government and oil company bureaucracy.

Since the advent of multiparty politics in Nigeria in 1999, I have participated in several visitations to the Niger Delta. The first was with the Ecumenical Council for Corporate Responsibility (ECCR), Catholic Relief

Services (CRS), and the Human Rights Commission (HRC)[4]. The second was with Trocaire, Catholic Relief Services, Collaborative Development Action (CDA) and Ireland Aid.[5] My third official visit was to Bonny Island[6] to help bail out a young man who the police detained over a debt issue. All these visits brought home the reality of the Niger Delta to my consciousness. As I write this I have worked on issues in and around the ND for more than ten consecutive years. I have also been involved as a conflict resolution practitioner particularly in the Niger Delta for sometime now. I have become deeply involved in the human rights movement. These are the 'baggage' that I am bringing into this project.

But a key challenge for me is that of switching my identity from an activist to that of a researcher. This is because as an activist I was clear about my mandate. I was on the side of the oppressed. And the oppressed in this case are people of ND. All the work I did was for them. I did everything to remain in their good book. I was on their side even when I disagreed with some of their strategies like hostage-taking which I can relate to as a former student activist.

Now that I am a researcher can I use all the privileged information, which I have had access to in my work? Or do I need to start afresh to gain the consent of my sources to use them? What do I do with such information that might endanger my sources even though they will enrich my work? After this project do I still have the moral right to go to the ND to work as an activist? Can I be trusted again? Will this work even be fair to the parties in the ND issue? Where can I draw the line as to where to begin and where to stop? These are some of the ethical issues that will haunt this project.

To come round all these I have decided on one thing. I am still on the side of the people of the ND and all the oppressed. This work is for them. It is to highlight their plight and to suggest new ways of transforming the conflict. But I will do all these by observing the finest of academic traditions of writing. The facts will remain sacred while my interpretations and opinions will be mine. I walk a mine but I think that my work in the human rights community has prepared me for the challenge ahead.

In chapter one I will deal with conceptual framework and definition of terms. Chapter two will locate the ND in Nigeria. It will also give a historical background of the conflict in the area and an overview of the ND. Chapter three will be literature review. This will examine what different people have to say about the ND problem. In chapter four I will make a theoretical statement and also look at the theories that will guide the main arguments of this project. Chapter five will analyze the responses of the government, oil companies, communities and non-governmental organizations (NGOs) to conflicts in the ND. The main aim will be to give an overview of the assumptions and motivations guiding these various responses. Chapter six will discuss the title of this book which is redefining conflict resolution initiatives in the Niger Delta region of Nigeria. Chapter seven will concentrate on my model for transforming conflicts in the ND. It will be a kind of strategic action plan for the ND. The emphasis here will be on how to use the model I developed called Community Education and Institution-building (CEIB) to transform the conflicts in the ND. Part of this model will be the use of what I will refer to as critical constituency (CC). Chapter eight will be my personal reflections on the conflicts in the ND. This will be against the backdrop of my training as a conflict resolution academic, human rights activist, an indigene of the Niger Delta and my experience as a practitioner in the field of conflict transformation and peacebuilding.

Methodology

I have been a practitioner in the ND. By this I mean that I have done human rights work in the ND. I have intervened in conflicts between various communities and oil companies and between communities and government agencies. I have also conducted several trainings and consultations in the ND. More importantly I have 'experienced' the ND. All these will come in handy. I will draw on these. I interviewed a cross section of activists, politicians, oil company workers, community leaders and academics. The main crux of the interview was to find out from them, what they think the issues in the ND are.

But my research question was a very simple one: why is it that in spite of all the huge investments on peace in the ND, conflicts and violence in the area seems to be increasing instead of decreasing? What should be done to reduce the number of, and the intensity of conflicts in the ND? Finally, what are the causes of the conflicts in the ND?

On the whole I interviewed about thirty people and had informal conversations with so many others. The interviews lasted for between ten to thirty minutes with each person. Some were recorded on audiotape while the rest were videotaped. During the informal conversations I made notes with the help of my research assistants[7]. The interviews were conducted in English language except for most of the women; they were conducted in their local languages with a smattering of English. I did not have any problems with translation. The interviews were conducted in the homes of the respondents and at times during public gatherings. I did not transcribe the tapes due to limitations of time. What I did was to listen to the tapes and watched the videos over and over again.

In the interviews I was not concerned with such broad causes of conflicts as economic, political or psychological factors. Rather, I focused on specific issues where the communities disagreed with either the oil companies or other communities, or even where communities disagreed internally. Most of my respondents were very reluctant to talk about inter-communal conflicts, though they acknowledged that it exists and needs to be addressed. My emphasis was for instance on such issues as land and how the issue of land caused conflict among individuals and communities. The questions were semi-structured but I asked follow-up questions to clarify issues. One of the difficult moments of the interviews was when the people have to take their time to tell the story of their plight in the hands of the oil companies.

A very surprising thing for me was the seeming non-existence of government in the responses of my respondents. For instance many of the respondents did not have anything to say about the government. When prodded, their typical response was, but "it is the oil companies we see, and it is them we know". Because of the sensitive nature of the conflicts in the ND, some of the respondents shall remain anonymous while others may be named depending on the kind of information they volunteered. Though they specifically did not request for this, I took the decision based on my previous experience working in the ND. I shall not mention all the cases, which I have been engaged in since some are still ongoing while the rest are in courts. But any reader interested in details of certain cases referred to in this book could contact me.

I did a lot of background reading especially newspapers, minutes of meetings and other publications. I also read books and articles not only on the ND but also on Nigeria in general. I consulted several reports posted on the web by activists, oil companies and other NGOs. But in using these sources I was very discerning because many were propaganda materials. I took extra care in analyzing them in order to pick as 'accurate' information as possible. By accurate I mean reconciling what is on paper with what I saw on ground and what I was told by my respondents. For instance in some oil companies' annual reports they give the impression that their expenditure on community development projects is on the increase. But what they do not say is that most of what is considered as community projects like roads is part of their normal business cost. The roads that oil companies allegedly built for the communities also provide access to their installations and sites.

Let me say a few things about my use of newspapers. I used newspapers only for information not for opinions. In the same vein I used NGO publications. The emphasis here again was on actual events to corroborate dates, figures and location of events. To a large extent the opinions are mine, which I formed after careful analyses of the data available to me. I also relied on my first hand experience working in the ND. For instance in the course of this research I joined a team from my office and we organized a meeting with Shell Petroleum Development Company (SPDC, but for this project I will simply refer to them as Shell). I conducted about three other trainings in the ND during the period of this research. All these information I used to draw some of my conclusions.

I conducted another round of visit to the ND (this time as a researcher, no longer as an activist) to seek clarifications on "matters arising". I participated in another stakeholders' review meeting of another oil company. Already I have had the rare privilege of participating in two of such meetings and I have both reports.

I also accessed the reports of shareholders and such bodies as the Ecumenical Council for Corporate Responsibility *(ECCR)* - a body that represents mainly church shareholders in most of these oil companies, Catholic Relief Services (CRS) and many others.

And then I also compared notes with those who have worked in the area and shared their insights. Finally I will apply some of the theories and models that I have been exposed to at the Conflict Transformation Program (CTP) of the Eastern Mennonite University (EMU), Virginia, USA to explain what I think is going on in the ND. Some classes, which I took at the Department of Conflict Analysis and Resolution of Nova Southeastern University, Florida, USA helped very much to put the issues into perspective.

However, there are limitations. Being a book for a general audience I could not be as detailed and as theoretical as I wanted to be. There is also the issue of space to consider so I did not use all the materials in my possession. In future I intend to upgrade and revise with some of these materials. Another limitation was that of time. During the period of research, I was also involved in many other commitments. So I may not have done justice to the work the way I would have wanted to. I was not able to interview all those I wanted to or visit the libraries etc. But I think that the materials, which I have, can really provide a good framework for starting. There is room for further work. This I intend to explore in future.

My being an Igbo[8] is also a limitation. There is this view within the ND communities that the Igbos oppressed them in the former Eastern Region. To a large extent this is understandable. So I am looked upon with suspicion. In fact I remember in one instance when I visited a community that had just experienced an oil spill, the first question a woman from the community asked me was "what is Abakaliki doing here?"

Abakaliki is one of the capital cities in Igbo heartland where my office is located. Now we have an office in Port Harcourt, one of the capital cities of the ND. I may not even be granted access to certain areas and information. Even the ones that will be made available may not be the whole pictures. I have worked in this area for sometime now. I have trained[9] a lot of people from the ND. Over time I have built up a network of relationships in the ND. I have also collaborated with a lot of activists from the area. And I have worked with people from the ND. I have participated in many projects in the area. All these I will tap into to write a scholarly and balanced analysis.

There can never be a final say on the issues of the ND. This is not one and is not even aspiring to be one. However, what I have can help me make the point, which is the main motivation for this work that we need to take a fresh look at what we have been doing in the ND with a view to rethinking our strategies for building sustainable peace and development.

My conclusion will not be the conventional one of summarizing the main arguments and findings of the research. Rather I shall use the conclusion to reflect on what I learnt working in the ND and researching this project. The last chapter will be my voice. I am doing this because like I said at the beginning, this project will not just be an academic excursion but a blueprint for intervention. This reflection shall be random thoughts and experiences or flashes of inspiration that I experienced in the course of doing this work.

Finally, in writing this I was guided by some basic assumptions, which I have developed over time working in the ND and researching this project. First, is that the more concessions the people of the Niger Delta get, the more likely they will escalate and sustain the intensity of the conflicts in the area. This is a strategy, which they have used over time. What then is the implication of this assumption for interventions in the area? Does it then mean that concessions should not be granted? Or that a hard-line position is the answer? My observation is that before addressing this question, we need to look at the so-called concessions and what they were meant to achieve. In this book, I have argued that the so-called concession did not have the people of the Niger Delta in mind. Second, due to the ineffectiveness of these concessions, the issues, which these concessions are meant to address, remain unaddressed. What then is the way out of this quagmire?

Second, laws have failed to mitigate or resolve the conflicts or even regulate the relationships in the ND. In this book I have demonstrated that the enactment of legislations is not the end of a conflict. First, the people must see the institution or government promulgating those legislations as legitimate. Second, those laws must been seen as just laws. Third, the institutions or government putting in place those legislations must have the will and institutional mechanisms to implement the laws. In the Niger Delta this has not been the case.

Third, coercion, force and intimidation can never solve the problem in the ND. In this book I have shown that almost all the parties in the conflicts in the Niger Delta have at one time or the other employed the use of violence to resolve their conflicts. The issue here is not whether the use of violence is right or wrong, even though for me my position is clear, which is that violence is not an effective conflict resolution mechanism. The issue is what are the motivations for the use of violence? What did the use of violence intend to achieve? What I found out from the interviews was that each party used violence or some other coercive or arm-twisting measures for different reasons. For instance, the communities used violence to draw attention to their plight. The government used violence to prevent the breakdown of law and order. Oil companies used violence to protect themselves from perceived threats. The implication of all these was that no party set out to use violence deliberately as a conflict resolution mechanism. Violence in most of the Niger Delta became the unintended consequences of a mismanaged response to a conflict situation.

Four, I think that the oil companies could with appropriate strategies and engagement at the right levels be ready to rebuild their relationships with the communities in the ND. My findings show that the oil companies are not the blood-sucking monsters that we have made of them. They have merely exploited the profit-motive to their own advantage. Second, they have also taken advantage of their relationship with government and especially the over-reliance of government on oil revenue to behave badly. But more importantly is that the human factor of incompetence, corruption, laziness and carelessness manifest in oil company operations. The strategic nature and the complex dynamics of oil industry operations are also some of the issues that have impacted their behaviour. The searchlight that has been beamed on oil companies cannot be divorced from the failure of government to be responsible and responsive.

Five, because of the excessive reliance on oil for revenue the governments in Nigeria cannot play the role of credible third party in the ND. The role of third parties in conflict intervention is well-documented in the literature. A critical factor in the Niger Delta has been the lack of credible third parties. For instance, the government could have played the role of an effective third party especially through such institutions as the legislature. But this is not the case because most Nigerian politicians need a foothold in the oil industry. A governor of one of the states in the Niger Delta has a business that relies heavily on the oil industry. There are even rumours that oil companies may have sponsored some politicians to their present position. On the other hand, the NGOs that would have acted as third parties are also struggling with issues of credibility and their history. For instance, the NGO movement in Nigeria rose and have survived based on opposition to governments. In fact, many NGOs cut their teeth under the military fighting for democracy. Unfortunately, in the current democratic dispensation, the military is still holding sway. So the government sees the NGOs as "opposition parties," while the oil companies worry about their credibility and capacity as third parties.

Six, only the communities can effectively define their relationship with the oil companies. While discussing the land issue I referred to theory of territorial imperative. The people of the Niger Delta own the land on which these resources are located. They also bear the brunt of its exploitation. They are in a position to define how the stranger in their land will behave. Unfortunately, with the advent of the modern state of Nigeria, the indigenous peoples of the Niger Delta seem to have lost this power. Part of the enduring legacy of the conflicts in the Niger Delta is that the government in far away Abuja (the capital of Nigeria) decides for the people of the Niger Delta how they will relate to a guest in their home. That is a problem. This situation is unacceptable to the people of the Niger Delta.

Seven, the mere citing of development projects (in this work I called it "Development Dumping") will not rid the ND of conflicts. Over the years the oil companies have moved from 'community assistance' to community development and now to sustainable community development. All these have been anchored around the misguided notion that development projects could be used to resolve conflicts. Though my findings suggest that the people of the Niger Delta are desirous of development in their locality, what has not been agreed upon is what their conception of development is. And what will be the modalities for the implementation of development projects. Very important is the fact that development, as Byrne observed must promote growth and change. This has not been the case in the Niger Delta. Development does not equal conflict resolution. A conflict must be resolved, then transformed and development comes within the integrated framework of peacebuilding.

Eight, that the ND communities must unite and speak with one voice before the rest of Nigeria will understand them. My interview with the elites and activists in the Niger Delta emphasize the need for unity, collaboration, coalition-building and synergy. This has been lacking in the Niger Delta. This is why I have argued in this book that the creation of states was not in the overall and long-term interest of the people of the Niger Delta. In another work, I have said that the creation of states fragmented the voice and struggle of the people of the Niger Delta. For instance, the rise of the Movement for the Survival of Ogoni People (MOSOP) under the mercurial leadership of Ken Saro Wiwa would have brought more and better dividends to the people of the Niger Delta if the efforts were better coordinated. It would have given them one voice and a collective vision.

The assumption is that creation of states would make more money available for the Niger Delta. First, the conflicts in the Niger Delta have little to do with money. Second, the creation of more states and local governments created new pockets of elites, bigger bureaucracy and democratized corruption. So little money is left for the execution of actual development projects in the Niger Delta.

Finally, that the solution to conflicts in the ND lies in sincere and facilitated dialogue between the government, the oil companies and the communities.

The content, modalities and dimension of this dialogue have been laid out in this book. Participants in this dialogue have also been suggested. The next thing is to start the process.

Notes and References

---

[1]Warri is the second largest oil-producing city in Nigeria.

[2] Because of politics we now have core and peripheral Niger Delta. Core refers to Bayelsa, Rivers and Delta States. But the Niger Delta is made up of nine states namely Abia, Akwa Ibom, Bayelsa, Cross River, Delta, Edo, Imo, Ondo and Rivers states.

[3] At independence in 1960, Nigeria had three regions namely East, West and North. Today she has thirty-six states.

[4] This visit was at the instance of ECCR (a faith-based NGO based in London) to monitor the extent of compliance to international standards by oil companies in Nigeria.

[5] This visit was to help Trocaire (development arm of the Catholic Bishops Conference of Ireland) define its intervention point in the Niger Delta.

[6] Bonny Island is the headquarters of the Liquefied Natural Gas (LNG) project.

[7] I had two research assistants, one guide and a cameraman. I am indeed grateful to them.

[8] The Igbos are the third largest ethnic group in Nigeria. And they were the dominant group in the former eastern region. The people of core ND were marginalized under the region. So they see the Igbos as part of the problem.

[9] I conducted trainings for ND and catholic youths in 2001 and student leaders in 2000 and nonviolence and peace-building. Some of the youths have provided me with access to their communities.

Chapter One

Conceptual Framework and Definition of Terms

There is need to clear doubts and misunderstandings about definitions and concepts. In this work, there are key concepts that need to be made contextually relevant. This is in order to understand how they are used in this project.

These concepts include the following: redefining, conflict resolution, initiatives and Niger Delta (ND). The ND we shall thoroughly examine in the next chapter. But in this work whenever we use the term ND we shall use it to represent the southeastern tip of Nigeria where oil is produced. Even though there is concerted effort to look for oil in other areas, for now the effort has not yielded any positive results. But in Nigeria's peculiar political parlance, there is the core and periphery ND. Rivers, Bayelsa and Delta states constitute the core while the periphery include Abia, Imo, Edo, Akwa-Ibom, Cross River and Ondo states. This qualification is simply due to the quantity of oil produced in each state. For instance, Bayelsa, Delta and Rivers states jointly produce 85.9% of total Nigeria crude oil.[1] Second, certain ethnic groups like the Ijaw, Itsekiri and Urhobo could be found more in the oil producing states; these ethnic groups are predominant in the 'core ND'.

Lewis Coser defines conflict as "a struggle over values and claims to scarce status, power and resources in which the aims of the opponents are to neutralize, injure or eliminate their rivals."[2] Embedded in this definition is the assumption that all conflicts lead to violence. At times what leads conflicting parties to try "to neutralize, injure or eliminate their rivals" is when the space has not been created for dialogue. The initial intentions in conflict situations are not to hurt someone. It is frustration at the inability to resolve or agree that leads to the desire to hurt the other party. In the ND, the people did not start out to hurt anyone. They wanted to better their lives. They wanted to partake in what God has given to them. They wanted to make sure that in the process of producing oil, that they are not destroyed through pollution, acid rain and lack of development. They did not want their land to become barren through oil operations. They did not want their rivers to become polluted so that they can no longer fish nor fetch water.

Coser also anchored his definition on the scarcity theory. This in other words means that conflict arises because resources are scarce. In the case of the ND this is scarcely the case. In the ND there is enough to go round. What has happened is that what is available has been mismanaged either through incompetence, greed or just mere ignorance. I agree with Coser that at times there could be struggle over values. But I think that the issue has more to do with recognizing and respecting other people's values than mere struggle. For instance, the oil companies tend to disregard the customary land rights of the people of the ND. By doing this the oil companies tend to devalue the customs and traditions of the people. This leads to conflict. Frynas refers to this as the ignorance and carelessness of oil companies[3]. Burton concurs with Coser when he defined conflict as including "struggles that are over resources, ideas, values, wishes, and deep-seated needs."[4] Burton introduced a new dimension to our understanding of conflicts by bringing in the issue of deep-seated needs. Rothman calls it identity-based conflicts. Identity issues have to do with who we are, what is core to us, our beliefs, values, culture, language, religion even space.

The issue here is that when identity is threatened, it could lead to conflict. In the Niger Delta this is true. This could be seen from the erosion of the culture of the people of the Niger Delta by oil company staff and the exploration of oil.

Hubert M. Blalock Jr. defines conflict as "the intentional mutual exchange of negative sanctions, or punitive behaviours, by two or more parties."[5] This definition like that of Coser tended to equate responses to conflicts to conflict itself. As we shall see in this book, the first response to conflict by the people of the Niger Delta is to write a letter seeking audience with the government or the oil company. It is when this fails or is frustrated that they consider other means. This has been the situation in the ND. According to Frynas, "the conflict between the family and the company escalated when Agip failed to respond to letters and representations from the local people"[6].

My concept of conflict in this context needs to be properly situated especially as it concerns the ND. Conflict is a social construction. In other words, conflicts do not just happen. They are created through deliberate human actions and interactions. It is also important to note that people do not just go out of their way to create conflicts.

However, it is people's perception, understanding and interpretation of events that makes them to decide whether an event is a conflict or not. This process of meaning-making comes from their culture. It is also shaped by their socialization and education process. All these have implications for how the people of the ND have responded to conflicts in their area.

Lederach captures this dynamic of conflict thus; "conflict is never a static phenomenon. It is expressive, dynamic, and dialectical in nature. Relationally based, conflict is born in the world of human meaning and perception. It is constantly changed by ongoing human interaction, and it continuously changes the very people who give it life and the social environment in which it is born, evolves, and perhaps ends."[7] This constitutes an apt description of the dynamics of conflicts in the Niger Delta.

More importantly, if we understand what their idea of conflicts is, then it helps us in the design of our intervention plan. For instance if training is part of our intervention plan, it would be very useful to incorporate their own indigenous method of conflict resolution as part of the curriculum. Moreover, it will also help us to know exactly the type of mechanism to use in resolving their conflicts. For instance some issues may have to go to court and others may be resolved outside the court system. A deep understanding of the people of the ND will help in all these.

My experience working in the ND for sometime now tells me that the people of the ND saw their conflict with the oil companies as injustice. They interpreted it in connection with power, that is if they were not minorities and powerless, that the oil companies would respect them. They made reference to other products in the other regions of the dominant ethnic groups. Here the people of the ND did not make a distinction between the actions and inactions of the Nigerian government and the carelessness and unresponsiveness of the oil companies to their complaints.

By injustice the people of the ND mean a deliberate act of oppression designed to put them down as a people, instead of seeing the issues in the area as a fall out of an interaction process that was mismanaged. In this line I argue that there is no grand conspiracy, what I saw and still seeing are only individuals or groups that are struggling to have access to resources like compensation for oil spill or for land acquisition. And in response to this perceived threat of extinction, the people of the ND fought to preserve themselves. At what point in time did they realize the reality of this threat? It goes far back in time, and this threat has been made real by the actions of the colonialists and various governments in Nigeria. But more importantly the exile of King Jaja of Opobo, killing of Isaac Adaka Boro and his compatriots and the judicial murder of Ken Saro Wiwa are all pointers to this theory of gradual liquidation.

As Docherty[8] noted, people bring their "gods" to the negotiation table. A key issue in a change process is what do the people hold dear? What is sacred and what is valuable? It is when we explore the worldviews of these communities that we may be in a position to really work with them. This is more so in the ND where most of the interventions have come from outside. By the interventions coming from outside, I mean that the oil companies based on their budget and policy decide what and what not to do in the ND. The government based on political expediency, budget constraints and power considerations decide the kind of intervention they want in the ND. The NGOs based on their mandate, funders' requirements and evaluation criteria also decide how and what to do in the ND. In view of this we need to understand what the ND was like before the advent of oil and project what it might be like when oil will no longer be the premier foreign exchange earner.

It is also worrisome that many interventions have failed to listen to what the people of the ND are saying. For instance, at the Human Rights Violation and Investigation Commission (HRVIC a.k.a Oputa panel), while Ogoni people were complaining of the desecration of their land, the panel was busy reconciling the factions within the Movement for the Survival of Ogoni People (MOSOP)[9]. Both Nigeria and the international community have never for once given the people of the ND the opportunity to say what they want without interruptions. With this closing of the space for dialogue, all we hear about are poorly articulated and unrealistic demands from faceless groups whose membership is as amorphous as their names sound.

Part of this nagging issue, which I intend to mainstream in this project, is why there seem to be too much mistrust and disunity among the people of the ND[10]. Divergence of opinion is normal in human society. What is abnormal however, is the inability to peacefully reconcile this divergence. Moreover any attempts to confront this distrust among the people of the ND have been resisted[11].

This also goes back to the issue of the desired outcome. If resource control is the issue or outright secession is the panacea, what are the guarantees that these conflicts will not resurrect in another form as pointed out by Kriesberg[12]. The inter-communal conflicts and the rivalry in the ND are all pointers that more work needs to be done before we can build sustainable peace. This is the main aim of the model CEIB that I am proposing in this work.

By initiative I mean any action taken to solve a problem or resolve a conflict. These include petition and letter writing, trainings, meetings, arbitration, mediation, third party consultation, facilitation, conciliation, demonstrations and protests, litigation and even silence[13]. Initiatives could be violent or non-violent. Needless emphasizing that I totally disagree with violent initiatives. Example of violent intervention is the Odi massacre of 1999, the Umuechem issue and the militarization of the ND generally. It is important to note that all the parties namely government, the communities and oil companies all have utilized violence as a form of response at one time or the other.

Nonviolent responses include the development project of oil companies, the establishment of such bodies as Niger Delta Development Commission (NDDC) and Oil Minerals Producing Areas

Development Commission (OMPADEC) by government and the other initiatives by various NGOs. The blistering media campaigns by the various parties in the ND could also constitute non-violent responses. However, non-violent actions require a lot of planning, painstakingness and creativity.

Responses to conflicts could be collective or individualistic, formal or informal. Initiatives are also targeted at a specific issue. But a key point to note is that initiatives are designed based on one's understanding of the issue to be addressed. A poor understanding of an issue will lead to an ineffective or poorly planned and executed initiative. Initiatives could be short or long term. When an intervention is short term, I shall refer to it as dispute settlement[14], when it is medium term I shall call it conflict resolution, but when it is long term and involves systemic change, I will call it conflict transformation. All these will become more explicit and clear as we progress in this discussion. The important thing therefore is that the above concepts are not mutually exclusive but mutually reinforcing. In other words they complement each other. The difference is not in kind but in degree.

The Webster's New World Dictionary explains the word define as[15] "to determine and set down the boundaries of", "to trace the precise outline of", to determine or state the extent and nature of; describe exactly or to give the distinguishing characteristics, to constitute the distinction of; differentiate or to state the meaning of" something. In this work I use the word to define, to mean, "To determine or state the extent and nature of" and "to trace the precise outline of" conflict resolution initiatives in the ND.

By using the prefix 're', I mean to re-examine the extent and nature of conflict resolution initiatives in the ND. My issue here is that people have already "determined the nature and extent of their own

31

initiatives". In these initiatives they have boxed themselves into a precise outline. For instance, the government is stuck with security concerns and the prevention of crime and "development dumping". The government, oil companies and communities have made up their minds as to what is the problem in the ND. With this they also came up with a solution or their initiatives. My challenge is to look at why these initiatives have failed to bring sustainable peace and development to the ND. I shall not reinvent the windmill that is why I am redefining. And this includes redefining the problems and the solutions. This redefinition will be in the form of analyzing what has been done with a view to critiquing it for lessons learnt and suggesting ways forward.

My redefinition of the issue in the ND is that it is conflict at various levels. In my research I identified eight different conflicts in the ND. And my redefinition of the solution is that it must be tackled at different levels to achieve the desired outcome. A key issue about the conflicts in the ND is what the desired outcome would be. For instance, the Ogoni at one time asked for self-determination, when confronted with the term secession they retracted their footsteps[16]. But again we may also need to agree on what the desired outcome should be. For instance if we ask for more money for the ND, is it more money for individual's pocket or for the communities or for the entire state for development purposes. Who defines this desired outcome becomes a contentious issue. But it would be helpful to work with some level of consensus rather than controversy. By integrated, I mean that a proper conflict analysis must be done, with appropriate models and an integrated multidimensional approach initiative must also be put in place. An entry point must be determined and specific actions taken. Exits must also be carefully planned. Effective monitoring and evaluation mechanisms must also be part of the initiative. A key issue will be how we put in place mechanisms for the nonviolent resolution of conflicts when they arise. By redefining I also mean that the government, oil companies, host communities and NGOs must collaborate to make the desired impact in the ND. This has not been happening, parties have been working at cross-purposes.

Finally let me quickly look at the term conflict resolution. Conflict Resolution was coined by Kenneth Boulding in 1957[17]. In an everyday sense, it means to solve a specific problem. It has embedded in it a short-term perspective. It is also shortsighted on relationship building[18]. It seems to ignore the expressive and often involving interactions between people in conflict situations. This is not to say that conflict resolution is bad. The issue is that what is happening in the ND is more than just solving a specific problem. It is easy to say for instance "pay compensation for oil spill." The question then arises as to whom should the compensation be paid to? How much should be paid? Should compensation be paid for damaged crops or structures on land or for land itself? It becomes more complex when one realizes that land is not all about earning a living but it is also a heritage bequeathed from generation to generation. If I do not accept the term conflict resolution as appropriate for what is happening in the ND, why have I implied such in the title of this work? I am using it because it easily resonates with people. What has happened in the ND has not been conflict resolution. Governments, communities and activists have always responded to issues in a generic manner.

There seem not to have been a holistic, consistent and sustainable approach to issues in the ND. Second, most of the initiatives to resolve the ND issues have not been altruistic and context-specific. They have either been politically motivated, for public relations purposes or downright dishonest. This is how Osaghae describes the creation of the Midwest region of 1963 for instance, "the region was created for reason which had more to do with the inter-party rivalry than any genuine desire to solve the Midwest minorities' problem."[19]

So my use of the term conflict resolution is to draw attention to the haphazard attempt at solving the ND problem. The isolated and selective reaction to issues in the ND represents only an attempt at conflict resolution. This is because even the so-called attempts have not succeeded in reducing the levels of violence in the ND. Conflicts are not only resolved, they are also transformed for sustainable peace. According to Byrne conflict transformation "focuses on how, what, and who transforms institutions, substantive issues, structures, and relationships to build a just and sustainable culture of peace."[20] In concluding this discussion I shall propose a new approach and a series of new terms. The purpose will be to get our concepts right and straighten our frameworks for analysis and intervention.

The main thrust of this book therefore will be to give a fresh framework for intervention in the ND. This will be done by pulling from the experiences and writings of practitioners and academics in the field. The main justification for this approach is because conflict is a dynamic process loaded with meaning[21]. Second is that we need to know where we are as interveners and what our limitations are in the intervention process. For instance, we cannot be working at the community level thinking or making promises that we are going to change oil prices. We need also to be aware of the resources available to us and our capabilities as individuals and organizations.

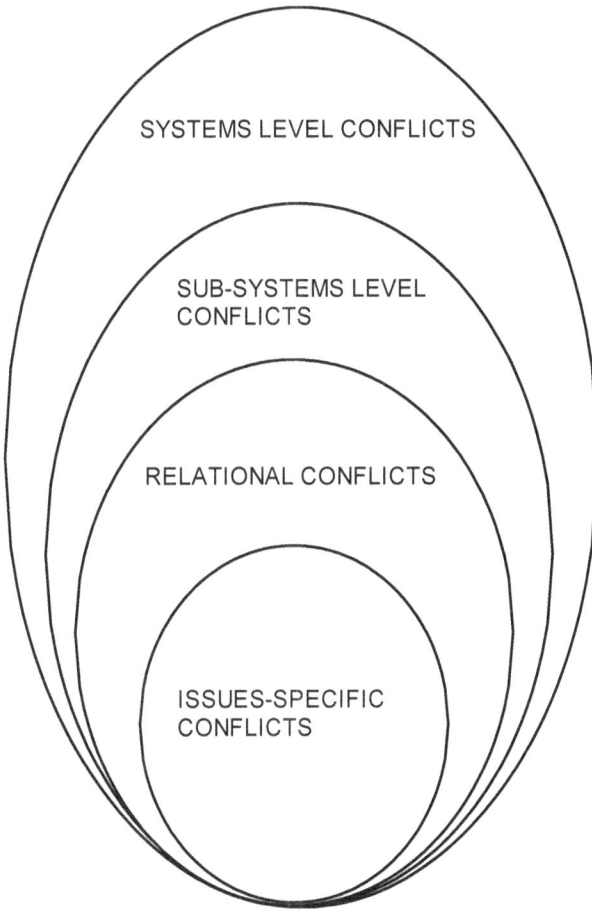

A Nested Theory of Conflict - courtesy of Maire Dugan.

Dugan[22] developed the "Nested Model" of understanding social conflict. In this model she identified four levels of analysis. The first is the system level, second is the sub-system while relational and issue-specific constitute the third and fourth levels respectively. Applying this to the ND, the system level is what happens at the international and national levels that affect what goes on in the ND. For instance the price of oil at the international market may make it impossible for government to provide the much needed amenities in the ND. In 1998 for example, the Nigerian government based the annual budget on an estimated price of US$17 per barrel, after the announcement of the budget, the price of crude oil dropped to US$14.73[23]. Second is that since most of the oil companies are multinational corporations, their policies are made outside the shores of Nigeria. Even where the price of oil rises, corruption, which is endemic and systemic, may make it impossible to improve the lot of the people of ND with the windfall[24].

At the sub-system level we have what goes on at the state and local government levels in the ND. For instance, if the state and local governments cannot effectively enforce environmental regulations, then it will affect the ND. On the other hand what happens at Shell headquarters in Nigeria may also be beyond the regional offices of Shell in the ND.

At the relational level, will be the relationship between the oil companies and their host communities. It may also be the relationship between the oil companies and the government. Specific issue here may be something like oil spill. This may involve the payment of inadequate compensation or boundary dispute between two communities. A good understanding of this is a *sine qua non* for effective intervention. From the eight different conflicts identified in the ND we can see that while some like oil spills could be relational and specific, others like land could be both relational and specific and sub-system and even systemic. There is the land tenure system as regulated by the Land Use Decree of 1978, there is also the people's own land tenure system and there is the specific issue of individual, family and communal land ownership. The challenge therefore for any intervener is to be clear at what level one is operating. Second is also to be clear as to the resources and skills required for intervention at the various levels.

Dugan's nested model is very important for understanding the conflicts in the ND because many interveners have been engaged in different things without knowing exactly what issues they are confronting. An example is the issue of human rights violations, which has become the battle cry for the conflicts in the ND. By releasing someone from police detention, that is addressing a specific human rights issue. But by engaging in legislative advocacy belongs to the realm of intervening at the system level. What one notices in the ND is that the interventions have been more at the issue-specific level. As we shall see in the later chapters, this does not augur well for the conflict transformation and peacebuilding paradigm, which is the case I am making in this book.

Working at the various levels of Dugan's model requires different timeframes. Specific issues may fall into a short timeframe, while systemic change might take a little bit of time. Lederach illustrated this in a diagrammatic form.

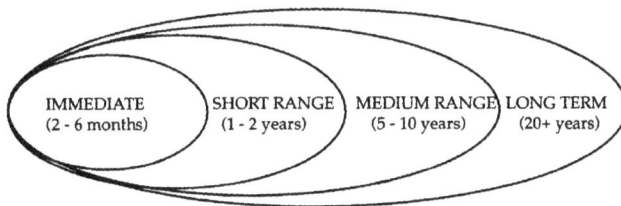

Time Dimension of Peacebuilding: Courtesy of Lederach (1999, 77).

An initiative may be long/short term. Lederach discussed this in his "Time Dimension in Peacebuilding"[25]. This gives an idea of the time required for each initiative. The first is the crisis intervention, which is between two and six months. This is related to Dugan's issue specific initiative, which is also of a short time limit. For instance in Dugan's and Lederach's models an issue like compensation should or ought to be resolved within a short time of about six months. In the ND this is not always the case. For example, it took Shell approximately one year to award the contract for the clean-up of the oil spill at Ogbodo in 2001. By September 2002, the issue of the clean-up was still in contention this time in far away London. Apart from this, the issue of whether the clean-up was properly done or not was still reverberating. Apart from this, the issue of compensation is still outstanding[26].

This time lag is what leads to frustration, which manifests in such issues as hostage-taking and the laying of siege on oil company installations. As we are going to argue, the oil spill is not the cause of the conflict per se, but the conflict arose out of the response of the oil company to the spill. And in intervening in the conflicts in the ND, it is important to bear this distinction in mind. This is because as long as oil exploration is going on, there will be spills, how these spills are managed becomes the driving points of our intervention. This insight has been missing in the analysis of the conflicts in the ND.

For instance negotiating the release of hostages and the stopping of violence or a siege are all short-term crisis interventions. But the issue of putting in place mechanisms for a timely response to the issue of oil spill and environmental degradation will take sometime. To my mind the efforts of interveners should be directed at preventing the situations that give rise to hostage-taking and incessant oil spills.

However, it is instructive to note that in the conflicts we identified in the research, each one could fall into Lederach's category. For instance, resolving land disputes take a little bit of time but the disputants could be given a reprieve within a specific time frame. I think that the main point of Lederach's model is the fact that interventions should not be a "touch and go" affair. This probably explains why in spite of the huge amounts invested in training peace seem to elude the ND.

One of the key lessons which we can learn from Lederach's time dimension of peacebuilding especially as it relates to the ND is that there have not been any long term peacebuilding effort that have crossed the five and ten years time limit. The implication of this is that interventions in the ND have hovered around crisis intervention and short-term conflict resolution. Even the government agencies that were supposed to permanently address the ND issue have often been short-lived either as a result of political instability or corruption. For instance, the Niger Delta Development Board, which was established in 1961, died naturally in 1966 after the coup. Oil Minerals Producing Areas Development Commission (OMPADEC), which was established in 1992, ran into trouble almost two years later. The Petroleum Special Trust Fund (PTF), which came into being in 1994, was disbanded in 1999. This also implies that most of these interventions have been lacking in symbolism.

Let me illustrate this issue of symbolism with the siege at Umuechem. The payment of compensation to the people of Umuechem should have been under crisis intervention, which falls into the two to six months timeframe. But it took the people of Umuechem almost two years to get the checks, and almost fourteen years after, the bounced checks are yet to be redeemed. Because of this, the healing, relationship-building and reconciliation which are all part of the long term intervention, cannot even start. Even if it starts, it may not succeed because the appropriate enabling environment has not been created. The symbolism of all these to the people of Umuechem is that no one cares in spite of their contribution to the national revenue stream. Whenever they look around them they see their burnt houses, they feel the loss of their loved ones and their polluted environment.

When a project is short term it lacks in-built mechanisms for sustainability. It will also affect the symbolism of land and the payment of compensation. For instance if one simply goes on acquiring and desecrating people's land because he can pay compensation, then not much will be achieved. For instance how much compensation will assuage a violated burial site or sacred grove?

So apart from the social and material world, Docherty has added the symbolic world[27] as another dimension to effective intervention and accurate understanding of the dimension, nature and dynamics of conflict. Docherty identified "three worlds" where conflict takes place. First is the symbolic world, second and third are the social and material worlds respectively.

The symbolic world represents the values that we bring into conflict and the meaning we attach to it. In the ND their rivers and streams are not just where they collect water for use but represent something deeper. For instance, among some people of the ND, when a child is born, he or she is taken to the river after eight days and initiated. So imagine a situation where the river where this child is to be initiated has been polluted by an oil spill. Imagine also that this is probably a community where it is believed that without this initiation that the child will become sick or die or be useless in life. All these will definitely complicate the conflict situation.

The values of the people of the ND include their land tenure system, which is part of their social system. The non-recognition of this customary land rights is an affront on the very fabric of their society. This also affects their relational world, which falls into Docherty's social world category. Docherty's social world falls into Dugan's relational issue. Therefore an upset in one world causes a ripple in the other worlds.

The material world, which has more to do with the issue of resources, is what has been emphasized in mainstream literature on the conflicts in the ND. This was done under the mistaken assumption that more money will de-escalate the intensity and number of conflicts in the area. This has not been the case. Moreover, as I shall show later, the paradox of the dynamics of conflict escalation is that the more concessions the parties get, the more the conflict intensifies. We shall show how the conflicts in the ND reflect this phenomenon.

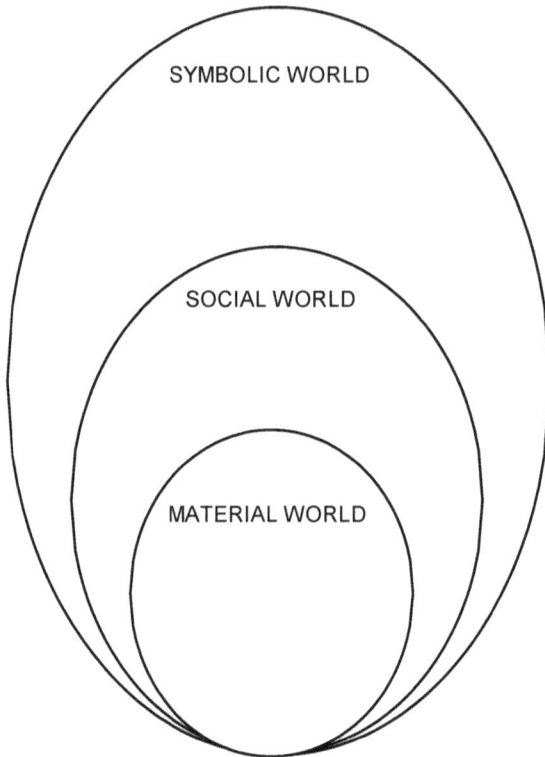

Three worlds where conflict takes place: Courtesy of Jayne Docherty

The ND issue has a lot of history to it. That is why for any initiative to be effective it must take into consideration Adam Curle's model of conflict progression[28]. The ND has fluctuated between peaceful and 'unpeaceful' situations. Curle explained it as part of the dynamics of conflict. For any initiative to achieve the desired result, then it must be able to understand the stage of the conflict and the premise on which to start the intervention. When we analyze the responses of NGOs to conflicts in the ND, we will see that they created a lot of awareness that escalated the conflicts in the ND. The same could be said about the various training programmes. After creating this awareness, the issue is what next? It is obvious that after a people are conscientised and made aware of their unjust situations, conflicts often escalate. Rothman[29] called it "antagonism" in his ARIA model while Curle illustrated this in his diagram. This has been done in the ND but without follow up on how to manage this new consciousness.

This may account for the increasing levels of violence and the intensity of the conflicts in the ND. For instance when Ken Saro Wiwa established MOSOP, it was not a secret cult but a mass movement that has clearly articulated goals with several publications. In fact the outings of MOSOP was what created ripples among Nigeria's political establishment. But little was done by the government and oil companies to constructively engage MOSOP. But today in its place we have ethnic militias that are faceless and ubiquitous without clear, achievable and measurable objectives. The import of this is that the new consciousness is being misdirected by people without the requisite skills for non-violent and peaceful change.

In the ND there is power asymmetry between the oil companies and the host communities. Even in negotiations and litigations this power disequilibrium is obvious. But the awareness of the people of the ND about the conflict in their area is high. This may probably explain the intensity of the conflicts in the area. However, the challenge is how do we balance power between the host communities, the oil companies and the government? This question is important because considering Nigeria's political arrangement, the ND constitute a minority. To my mind, balancing power in this instance is out of the question. The issue will be empowering the people with skills to constructively engage the oil companies and government. This will have more to do with understanding the oil industry.

	Unpeaceful	**RELATIONS**	Peaceful
	Static	**Unstable**	**Dynamic**

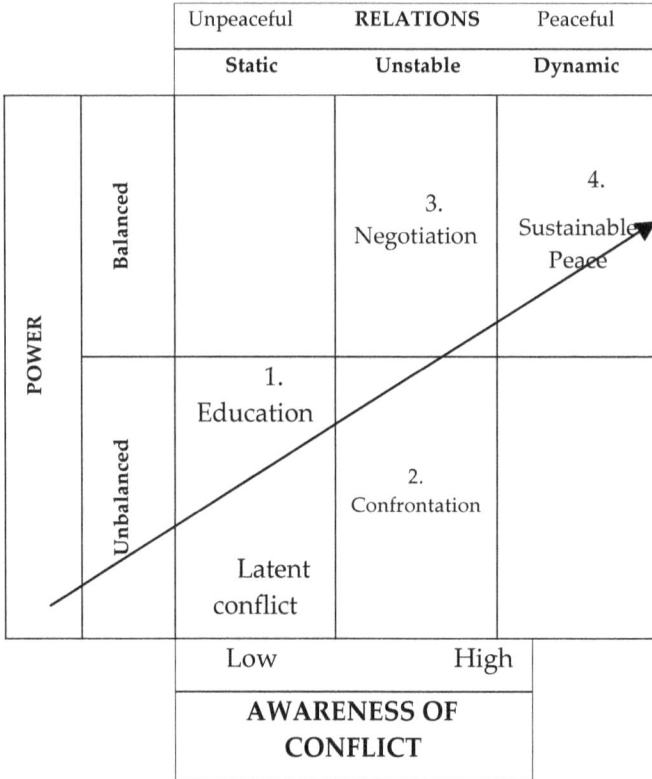

Source: Adapted from Curle (1971) and Lederach (1997)

More importantly the significance of Curle's model, when applied to the ND is that time is required for sustainable peacebuilding efforts. Second is that this will be done in phases. Third is that every stage will have certain basic structures to be put in place. For instance, at stage where the relationship will become peaceful through such tools as negotiation and mediation, training will need to be conducted for the people to learn to negotiate with people that are coming from different backgrounds and culture. For instance, in a negotiation session with an oil company, the community may have to be made aware of the short term and profit motive of oil companies, while the community will be coming from a premise of long term and existential mode. Reconciling these is part of the intervention.

For a better understanding of the parties involved in the conflicts in the ND let us look at what Lederach calls The Pyramid Model of Actors[30]. In this model Lederach identified three levels of actors namely Track One, Two and Three. Track One includes top politicians, military and religious leaders etc. Track Two is the middle people that include credible and nondescript leaders such as retired teachers, NGO workers, local religious leaders etc. Track Three is the grassroots people. These are those that bear the brunt of the effect of the conflicts. These include community leaders, women's groups, youth associations, age grades etc. The essence of all these is to know who to work with in designing the initiative and who the intervention will impact most. It will also help us to identify and recruit allies and identify and possibly convert spoilers.

In the ND there has not been a carefully articulated process of engagement. For instance, I have been engaged in several training programmes in the ND, not once have I seen law enforcement agents, oil company workers and the host communities sit in the same room. The implication of this is that each group talks past each other. In fact, the oil companies are so suspicious of NGOs that they will do anything to get them out of the way. While the oil companies accuse NGOs of lacking in integrity, NGOs accuse the oil companies of lack of transparency in their dealings with the communities. The government would rather have nothing to do with the NGOs because according to them they are troublemakers. There is a need for shared visions for the ND between the parties especially for sustainable peace.

Actors and Approaches to Peacebuilding
Actors at different levels play different roles in both conflict and peacebuilding.
J.P. Lederach, p. 39 (Building Peace)

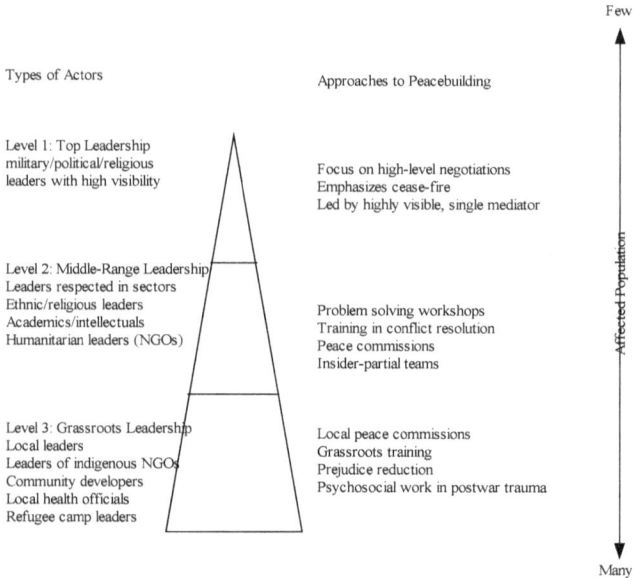

Few

Types of Actors

Approaches to Peacebuilding

Level 1: Top Leadership
military/political/religious
leaders with high visibility

Focus on high-level negotiations
Emphasizes cease-fire
Led by highly visible, single mediator

Level 2: Middle-Range Leadership
Leaders respected in sectors
Ethnic/religious leaders
Academics/intellectuals
Humanitarian leaders (NGOs)

Problem solving workshops
Training in conflict resolution
Peace commissions
Insider-partial teams

Level 3: Grassroots Leadership
Local leaders
Leaders of indigenous NGOs
Community developers
Local health officials
Refugee camp leaders

Local peace commissions
Grassroots training
Prejudice reduction
Psychosocial work in postwar trauma

Affected Population

Many

There are two ways to understand this Pyramid Model of Actors. First, is that we can use it to understand the various parties involved in the conflict. Second it can also help us to understand the various levels of intervention and interveners. As we have in the ND, the people most affected by the conflict are people at the Track Three. And most of the interventions have been at that level. The implication of this is that the whole effort has been to sort of suppress the conflict. But this has inadvertently intensified the conflicts and brought it into the open. The fact of the matter is that as far as conflicts in the ND is concerned; Track One actors have not been engaged. What may pass as engagement could be termed the war of words or the political opposition posturing of interveners. This shall be carefully analyzed when we discuss responses.

The important point to note here is how do we then design and manage an intervention package that will cut across all the three levels of actors and also impact all of them. What we see in the ND especially among Track Two people is the demonisation of every one in Track One. The track one people are seen as enemies that should not be hobnobbed with. If we are to achieve sustainable peace and development in the ND, then all the actors must join together. We must also have interveners that have the required skills and training to feel at home at all three levels. This is not the case as at now.

Still on the actors, is the seeming confusion as to who the parties in the conflicts are. We know that the stakeholders in the ND are the communities, government and the oil companies. The NGOs come in more as interveners who are also stakeholders. But there is a problem when it comes to who is in conflict with whom. The various communal conflicts in the ND pitch the people against each other. The fallouts of oil exploration activities pit the people against the oil companies. The lack of infrastructure pitches the people against the government. This is very complex especially in a federal state like Nigeria where it is difficult to know who is responsible for what. So it is very difficult to know who is engaging who and for what and why?

Another point to note in this engagement process is that conflicts in the ND are not about to end. Even if oil dries up today conflicts in the ND will not go away. A pointer to this fact is the conflicts all over the world including various parts of Nigeria. Many interveners have come with the mistaken assumption that once they intervene the conflicts in the area will end. It is better to be informed by Edward Azar's[31] theory of Protracted Social Conflict. Coleman calls it intractable conflicts[32] while Rothman[33] calls it intransigent conflicts. The characteristics of these conflicts as Coleman noted are that they have "an extensive past, turbulent present and a murky future". Second, these conflicts include both tangible (resource) and intangible (identity) issues. The issues, parties and contexts are constantly evolving. They are also tied to John Burton's[34] human needs of security, safety, dignity and belongingness.

Lending his voice Kriesberg[35] opines that certain conflicts are described as social conflicts because they involve interaction between individuals or groups. And that as long as these people interact that there will always be conflict. The issue then is that conflict will manifest in different ways in that relationship. Azar admonishes us not to deceive ourselves that the conflicts will finally be resolved once and for all. He says that the better way to look at our intervention is to put in place mechanisms for long-term conflict transformation. This brings us back to the issue of our integrated approach. So for a successful intervention in the ND, while working at specific relational issue, we must also do long term conflict transformation which according to Curle involves education, action and advocacy.

Though protracted social conflicts may not be resolved once and for all, we can at least minimize both the perception and actual deprivation among the people of the ND. And Gurr's theory of relative deprivation[36] may provide the road map for a better understanding of the conflicts in the ND. Relative deprivation is a simple theory that says that the more people are deprived of what they consider their due, against what their compatriots are getting, that they are likely to rebel. In Gurr's words "relative deprivation is a perceived discrepancy between men's value expectations and their value capabilities"[37]. Continuing, Gurr defines value expectations as "the goods and conditions of life to which people believe they are rightly entitled", while value capabilities are the goods and conditions they think they are capable of attaining or maintaining given the social means available to them". To underscore the meaning and his use of the word relative, Gurr quoted Karl Marx and Friedrich Engels from their work *Wage Labour and Capital*, "our desires and pleasures spring from society; we measure them, therefore, by society and not by the objects which serve their satisfaction. Because they are of a social nature, they are of a relative nature"[38].

First, the import of this theory for the Niger Delta people is that they are comparing "the goods and conditions of life to which people believe they are rightly entitled" (value expectations) to the other ethnic groups in Nigeria.[39] For instance, when they compare the available infrastructure in their area to that in the other area, they are moved to protest. They also compare the number of their people employed in the oil companies with others from the other ethnic groups in Nigeria, and feel cheated. When they also compare contracts awarded to their people and others they feel betrayed by the Nigerian nation.

In the 1999 constitution of the Federal Republic of Nigeria, it was stated that 13% of total oil revenue should be reserved for developing the Niger Delta area. The Niger Delta people derisively refer to this provision as "87% deprivation"[40].

Clearly they feel that comparatively they are not getting enough and this according to Gurr could be a valid explanation for the escalation of violence in the area.

Another very interesting aspect of Gurr's theory is that it links the level of deprivation to that of participation. He identifies three levels of deprivation namely: mild deprivation, moderate and intense deprivations. In his own words "mild deprivation will motivate few to violence, moderate deprivation will push more across the threshold, very intense deprivation is likely to galvanize large segments of a political community into action"[41]. This is very true of the Niger Delta situation. In the 1960s and 70s when Nigerian economy was booming, there was muted protest in the Niger Delta but since the economy nosedived the conflict there has taken on a life of its own and almost everyone is involved.

On the other hand, if this Gurr's postulation is subjected to interrogation, we will notice some discrepancy especially with the situation in the ND. It was when Ken Saro Wiwa took over the leadership of MOSOP and brought his immense talent to bear on the movement that Ogoni people were mobilized. And MOSOP seems to have been the first mass movement that clearly articulated the issues of the ND. To what extent that MOSOP success could be attributed to increased deprivation is a topic for another research. But suffice to mention that in the case of the ND, leadership and organizational skills could account more for mass participation than increased deprivation.

Gurr also identified four other variables that may determine the propensity for violence. They are anticipated gain, opportunity, fear of retribution and legitimacy of government[42]. In the Niger Delta, the "anticipated gain" is to gain more access to the oil revenue. This is not only realistic but also attainable and this could be discerned with the kind of concessions that each regime makes to the Niger Delta. For instance, the establishment of OMPADEC, PTF, and NDDC are all efforts ostensibly by the government to address the issue of the ND. The various tinkering with the revenue allocation formula is also another example of this anticipated gain.

A critical study of Niger Delta uprisings will show that they have always taken advantage of political or other unrests in Nigeria to draw attention to their plight. It was in 1966 that Isaac Adaka Boro took advantage of the build-up of events that led to the civil war to declare the Niger Delta Republic. The June 12, 1993 election debacle provided another opportunity. They have also looked at the legitimacy of government and its ability to punish before they strike. The people of the ND have also taken advantage of the resurgence of human rights activism and the wave of democratization to amplify their agitation.

Continuing with his analysis, Gurr identified three stages in the development and actualization of political discontent. First, he talked about the development of discontent. This is the stage of the articulation of the grievance. It also involves the mapping out of strategies for agitation. The second stage is the "politicization of the discontent". In the Niger Delta, the protest against oil companies and government did not happen in one day. Initially whenever people had problems with the oil companies they settled it individually or directly or with the help of a lawyer. But gradually the issue was put on the agenda of Nigeria's political discourse. The establishment of MOSOP and such other groups are clear indicators of what Gurr is referring to. The process of actualization is an ongoing one at different levels and at different times. Recently, a group of Ijaw youths declared their intention to secede from Nigeria.

From our survey of various conflict interventions in the ND, it is obvious that many interveners were not informed by Talcott Parsons[43] Functionalist theory. Parsons compared human society to a biological organism. According to him the institutions of society are like the various parts of the human body. And each of these is related to one another. Each institution performs a role that makes for societal harmony and stability. He identified four functions of the various institutions of society. They are adaptation, goal attainment, integration and latency. Within this functionalist theory, he developed the concept of "homeostatic equilibrium". This concept simply states that if an institution changes, others must change to ensure stability and order.

Parsons is "stability" theory. How can it find relevance in a conflict context especially in the ND? To my mind Talcot Parson's functionalist theory is very important especially in designing an intervention plan. This is because whatever we do, we must be ready for the unintended consequences of our intervention. For instance if more money is allocated for developments projects in the Niger Delta, how will those deprived feel? If the oil companies vote more money for community development what will be the impact on their profit, the value of their stocks and the reaction of shareholders. So while it might be necessary to address relative deprivation, it is also important to look at how it is going to impact other variables.

First, we need to look at Nigeria as an organism, in which the Niger Delta is a part of and the oil industry is also a part. Second, the oil industry is also an institution within the global system. It is also a part of such body as Organization of Petroleum Exporting Countries (OPEC). Most of the oil companies are part of multinational corporations. Apart from this the Nigerian government does not solely determine what happens in the oil industry. Events within and outside Nigeria combine to impact the oil industry as well.

At the political level, the Niger Delta is part of the federal system of Nigeria. Anything happening there affects other parts of Nigeria. On the economic level, Nigeria, the oil industry and multinational corporations are part and parcel of the global economic system. For instance, Nigeria cannot just make laws regulating the industry indiscriminately without taking a cue from such organizations as OPEC and its members and home country governments of the multinational oil corporations in Nigeria. As they say, the world has become a global village.

At the social level, other ethnic groups cannot see government action as always appeasing the people of the Niger Delta. This may breed unrest in other parts of the federation. For example, the creation of states in one part of Nigeria has always led to more agitations in other parts. This is one of the reasons why government response to conflict in the ND has been heavy handed. And on the individual basis, government must ensure that it distributes resources equitably to as many members of the polity as possible. Parson suggests the concept of integration for this task.

The next issue is the implications of the above for practice. First, is that people must not be deprived. And where deprivation has taken place as in the Niger Delta institutional mechanisms must be put in place to redress this without depriving another group. So in this instance, the intervention must involve systemic issues. Since deprivation is relative, our intervention must not be "a rush to dump development projects". For instance, the oil companies have tried building hospitals, schools, sinking boreholes etc. all to no avail. So our intervention must be authentic, systematic, creative and original.

Our intervention must not be an isolated event. We must recognize as Parsons did the interdependence and interrelatedness of various institutions bearing in mind that whatever we do in one institution affects the other in one way or the other.

We must also anticipate the unintended consequences of our intervention. Since change at times can be painful, some of our intervention may reverberate in other areas we are not working. So we have to be prepared for this. The theories (especially Gurr's) emphasize the need to be pro-active instead of reactive in our intervention. For instance, the opportunity must not be created that will motivate people to rebel. And we must at all times ensure that we put mechanisms in place to monitor and evaluate our intervention efforts. It is by doing this we can know the impact of our actions on other parts of the "body".

What are the issues, lessons and insights that have been raised from the above redefinition of conflicts in the ND? The most striking for me is that interventions in the ND have been haphazard and uncoordinated. The reason for this as I explained in the opening chapter is probably because conflict resolution as a discipline with its own body of knowledge is still a relatively new discipline in Nigeria. It is also possible that people did not see what is going on in the ND as conflict that could be transformed. The issues of the ND have been seen as part of the political evolution of the Nigerian nation or part of the history of the ND[44]. I also think that poor analysis of the situation may also account for this ineffective intervention.

Most oil companies in the ND do not have a conflict resolution department. What they have mostly are community development desks. More importantly they argue that they are an oil exploration company and not a community relations or development outfit.

Another possible explanation is that interveners have tried to do every thing at the same time. Most NGOs do conflict resolution, they do human rights, they do development, they do environmental issues, they are into community development, political mobilization etc. All these become very confusing. This is not to deny the interrelatedness of these whole issues. The point however is that interveners must target specific issues and ensure that they transform them effectively.

I strongly feel that a lack of follow up on activities or intervention could also explain the failure of most interventions. For instance when I was conducting the interviews for this project, I asked an oil company executive to show me one community where they think that they have done a good job. His reply was as shocking as it was disappointing. He mentioned all the communities where we have been working in the last three years. Whether he was being sincere or ill-informed or downright mischievous I may never know. But his reply was very instructive.

In another instance, an employee of another oil company was in a meeting with me. I asked her what the situation in the ND was. She replied that they know what the NGOs are doing, that they are collecting money from abroad and pocketing same claiming that there is conflict in the ND. What this has to do with the situation in the ND still baffles me as I pen these words.

One of the issues that constantly erupted when I was conducting the research for this project is that of lack of respect for the people by the oil companies. In my work in the ND since 1999, I have come across these several times. For instance, when we were intervening in an oil spill situation, the oil company refused the community members to be present during the negotiations. When we also visited a government office in the ND, this same scene was re-enacted when government officials supposedly elected by the ND people refused to see them but was willing to see us from civil society. This probably explains why some oil companies hardly reply to letters from the communities.

The next issue is what does this redefinition of conflicts and interventions in the ND tell us about the work that has gone on in the ND? The first word that comes to my mind is confusion. I think that we cannot say with all amount of sincerity that there has been any meaningful intervention in the ND. This may sound as an overstatement but it is fair considering the fact that the situation in the area seems to be deteriorating. The above assertions may not be validated due to lack of such indicators as accurate statistics. This is not peculiar to the ND, it is a national malaise. This may also explain why there have been little or no methodological analyses of the conflicts in the ND.

It is also important to begin to document the impact of the conflicts in the ND. Even though statistics do not bleed but when people do, statistics helps us to feel the impact of the bleeding. I think that there is too much cover up of the conflicts in the ND, which has made it almost impossible to properly investigate the issues with a view to putting it on the national agenda. The first time this opportunity showed up at the Oputa panel it was mismanaged.

Notes and References

---

[1] OMPADEC Quarterly Report 1993, cited by Omotoye Olorode in Boiling Point 2000, p. 10.

[2] Coser, Lewis A. *The Functions of Social Conflict*. New York: Free Press, 1956.

[3] Frynas, Jedrzej Georg. *Oil in Nigeria, Conflict and Litigation between oil Companies and Village Communities*. Hamburg: LIT, 2000, p.180.

[4] Burton, J.W. *Conflict Resolution: Its language and Processes. 1996.*

[5] Blalock, H.M. Jr. Power and Conflict: *Toward a General Theory*. Newbury Park, CA: Sage Publications, 1989, p.7.

[6] Frynas, J.G. "Oil in Nigeria", p. 187.

[7] Lederach, J.P. *Building Peace: Sustainable Reconciliation in Divided Societies.* Washington DC: USIP, 1999, pp. 63 and 64.

[8] Docherty, J. S. Learning Lessons from Waco: *When the Parties Bring Their Gods to the Negotiation Table*. New York: Syracuse University Press, 2001.

[9] Report of Oputa Panel, 2002.

[10] See Osaghae, E.E. "Ogoni Uprising". P.334.

[11] Thisday, Sunday, January 21, 2001.

[12] Kriesberg, L. *Constructive Conflicts: From Escalation to Resolution.* Maryland: Rowman and Littlefield Publshers, 2003.

[13] Op. cit. p.106.

[14] See Brad Spangler, "Settlement, Resolution, Management and Transformation: An Explanation of Terms." (www.beyondintractability.org/m/meaning_resolution.jsp)

[15] Neufeldt, Victoria (Ed). *Webster's New World Dictionary of American English* (Third College Edition). Prentince Hall, New York, 1994.

[16] Osaghae, E. E. "The Ogoni Uprising: Oil Politics, Minority Agitation and the Future of the Nigerian State." *African Affairs,* 94, 1995. pp. 325-344.

[17]Woodhouse, Tom and Oliver R. (eds). *Peacekeeping and Conflict Resolution.* *Portland,* Oregon: Frank Cass Publishers, 2000, p.4.

[18] Christopher Mitchell disagrees with this notion. For details of his argument see Christopher Mitchell, "Beyond Resolution: What does conflict transformation actually transform?" Peace and Conflict Studies, Vol. 9, No. 1, May 2002, pp. 1-18.

[19] Osaghae, E. E. "Ethnic Minorities and Federalism in Nigeria." *African Affairs,* 90, 1991, p.243.

[20] Byrne, Sean. "Transformational Conflict Resolution and the Northern Ireland Conflict." IJWP, 2001.

[21] Lederach, J.P. *Preparing for Peace: Conflict Transformation Across Cultures.* New York: Syracuse University Press, 1995, p.8.

[22] Dugan, Maire A. "A Nested Theory of Conflict". Women in Leadership: Sharing the Vision, 1 July, 1996, pp. 9-20.

[23] *Newswatch* Magazine, February 16, 1998.

[24] Okonta, Ike and Oronto Douglas. *Where Vultures Feast: Shell, Human Rights and Oil in the Niger Delta.* San Francisco, CA: Sierra Club Books, 2001, p.37.

25 Lederach, J.P. *Building Peace: Sustainable Reconciliation in Divided Societies.* Washington, DC: USIP Press, 1999, p.77

26 The Center for Social and Corporate Responsibility has been on this case. I was also involved in this case as far back as 2001. See also "Extractive Industries Advocacy and Corporate Responsibility: case study of CSCR by Drs. E.O Emmanuel and David C. Okwudili , 2004.

27Docherty, J.S. Learning Lessons from Waco: *When the Parties Bring their Gods to the Negotiation Table.* Syracuse University Press: New York, 2001, p.31.

28 Curle, Adam. *Making Peace.* London: Tavistock Press, 1971.

29 Rothman, J. *Resolving Identity-based Conflict in Nations, Organisations and Communities.* San Francisco, CA: Jossey-Bass, 1997.

30 Lederach, J.P. *Building Peace: Sustainable Reconciliation in Divided Societies.* Washington, DC: USIP Press, 1999, p.40.

31 Azar, Edward. *The Management of Protracted Social Conflict.* Hampshire, England: Dartmouth, 1990.

32 Peter Coleman, "Intractable Conflict" in Deutsch, M and Peter Coleman. (Eds.) *The Handbook of Conflict Resolution: Theory and Practice.* San Francisco: Jossey-Bass, 2000, pp. 428-450.

33 Rothman, J. *Resolving Identity-based Conflict In Nations Organisations and communities.*

34 Burton, J.W. (ed.) *Conflict: Human Needs Theory.* New York: St. Martin's Press, 1990.

35 Kriesberg, L. *Constructive Conflicts: From Escalation to Resolution.* Maryland: Rowman and Littlefield Publishers, 2003, pp.2-24.

36 See Gurr, T.R. (1970). *Why Men Rebel.* Princeton, NJ: Princeton University Press.

37 Ibid: p.13

38 ibid: p.22

[39] See Patrick Fregene's lecture "How Nigeria Plundered and Underdeveloped the Itsekiri People" in Boiling Point cited earlier, pp. 111-132.

[40] I heard this from a participant in one of my trainings in the ND in 2000.

[41] See Gurr's Why Men Rebel, cited earlier.

[42] Ibid: p.9

[43]So, Alvin Y. *Social Change and Development: Modernization, Dependency and World Systems Theories*. Newbury Park, CA: Sage Publications, 1990.

[44] See for instance Prof. Julius Ihonvbere's views in Boiling Point, p.74.

Chapter Two

Background History of Nigeria

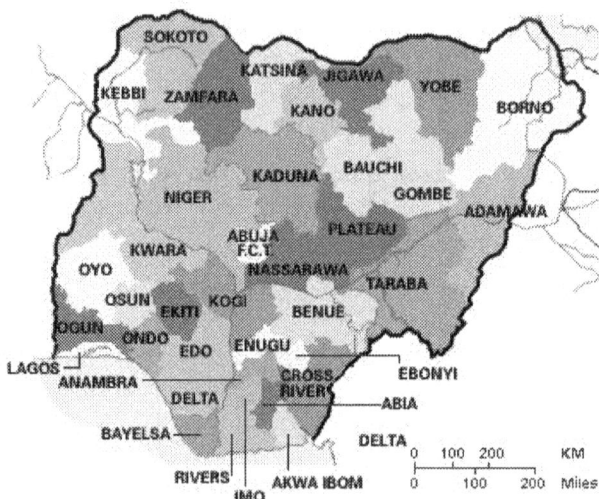

Map of Nigeria showing the 36 states. Courtesy of Shell Nigeria.

Nigeria is a contraption of convenience. Everything about Nigeria - ranging from name to political system was all due to convenience, not for altruism or systematic planning or symbolism. For instance, the name Nigeria was coined in 1897 by Flora Shaw, a correspondent of Times of London. The name was then adopted by the British in 1900. This is how Shaw put it:

> In the first place as the title Royal Niger Company's Territories is not only inconvenient to use but to some extent also misleading, it may be permissible to coin a shorter title for the agglomeration of pagan and Mohammedan States which have been brought by the exertions of the Royal Niger Company within the confines of a British Protectorate and thus need, for the first time in

60

history, to be described as an entity by some general name. To speak of them as the Central Sudan, which is the title accorded by some geographers and travellers, has the disadvantage of ignoring political frontier lines, while the word Sudan is too apt to connect itself in the public mind with the French Hinterland of Algeria, or the vexed question of the Nile basin. The name Nigeria applying to no other part of Africa, may, without offence to any neighbours, be accepted as co-extensive with the territories over which the Royal Niger company has extended British influence, and may serve to differentiate them equally from the British colonies of Lagos and the Niger Protectorate on the coast and from the French territories of the Upper Niger.[1]

Nigeria covers about 336,669 square miles and between 3 and 15 degrees east longitude and between 4 and 14 degrees North latitude. Nigeria stretches about 700 miles from east to west and 650 miles north to south. It is three times the size of its colonial master Britain, and twice the size of California. Nigeria is bounded in the north by Chad and Niger, west by Benin and Togo and east by Cameroon. The southern part of Nigeria is bounded by the Atlantic Ocean. Its two major rivers are the Niger and Benue. The River Niger, which is the third longest river in Africa, is about 730 miles across Nigeria[2].

Nigeria has about 250 ethnic groups with a population of about 130 million[3]. There were various levels of interactions between the various ethnic groups before the coming of the Europeans and the Arabs. But on the whole the ethnic groups remained independent and maintained their sovereignty. Interactions between them varied from inter-marriages to war to trade to diplomacy[4]. About 21% of the population speak Hausa, Yoruba 20%, Igbo 17% and 7% Fulani. English is the language of official business.

The vegetation varies from place to place. The north is mainly grassland and savannah. For a large portion of the year, there is little rain. The north experiences drought from time to time. The main crops of the north are groundnut, sorghum, corn, millet and wheat. The south has mainly dense rain forest and rivers, streams and creeks. The main crops are palm kernel and oil, cocoa, rubber and timber. The south has rain for most part of the year and is humid.

Production of most of the crops from the regions witnessed terminal declines from the time oil was discovered in 1956 at Oloibiri in the present day Niger Delta of Nigeria. Apart from oil, Nigeria is also endowed with other minerals. For instance, there is tin, uranium and columbite in the Middle Belt. There is coal in Enugu, salt in Abakaliki, limestone in Sokoto and zinc and iron ore in other areas.

Pan-African scholars[5] will object to one starting the story of Nigeria or that of any other African nation for that matter with a narrative of the activities of foreigners. According to these scholars, it gives the impression that the people have no history. They would rather see the activities of the colonialists as the main cause of distorted development in Africa. In other words, they argue that what the white man left in Africa was a trail of tears, sorrow and blood. Though to a large extent true, this should not make us lose sight of the fact that Africans have also contributed to their own misfortune. For instance, the carpet-crossing incident in the Western House of Assembly was engineered by a Nigerian politician. This signalled the demise of national political parties in Nigeria. Second, the annulment of the June 12, 1993 election won by M.K.O. Abiola (a southerner) by the Nigerian military also reinforced the retreat into ethnicity.

On the other hand, European and North American scholars[6] will argue that the colonialism argument is a worn-out one and that Africans must, and should begin to take responsibility for their development maladies. Both arguments have their merits, depending on how one looks at it.

First, Nigeria and most part of Africa is a victim of what Ali Mazrui[7] refers to as the "triple heritage". These heritages include Islam, European and African civilizations. This development has a profound effect on Nigeria especially as it concerns the conflicts in the Niger Delta. This will become apparent as this discussion unfolds. To the average Niger Delta person, Islam is the religion of those who are oppressing them today and carting away their God-given wealth. Europe represents the center of the slave trade, denial of participation in "legitimate trade" and worst still; the oil companies that the people of the Niger Delta have come to see as symbolizing their misfortune all have their roots in Europe and North America.

The Nigerian State as we know it today is a creation of the British. So there is no way one can discuss Nigeria without reference to the activities of the British. Second, what is more relevant to this study is that the root causes of the conflicts in the Niger Delta was laid at this time through a combination of social, political and constitutional mechanisms. Most of the laws and policies[8] made during the colonial period have either been updated, or reviewed but never totally discarded[9]. More so, Crawford Young has argued that "the unending turbulence in this region provides daily confirmation of the colonial roots of many intractable contemporary conflicts"[10].

On the other hand, before the amalgamation, the north had in fact been conquered by Muslims. So while Europeans were rampaging the south, Arabs were consolidating their conquest of the north. The British did not help matters either when they eventually made inroads into the north. They decided to minimize contacts between the northerners and southerners[11]. This was because they saw the southerners who had acquired western education as troublemakers because they were in the forefront of the nationalist movement. This led to so many concessions[12] being granted to the north in the various constitutional conferences. For instance, in 1939, the Eastern and Western regions were created from the old Southern protectorate but the North remained untouched. Second, in 1947, the Richard's Constitution created a Central legislature for Nigeria. The same constitution also made provision for regional councils thereby protecting the north from 'contamination' from the south. It was at one of these constitutional conferences in 1952 that the North insisted and got 50% of the national legislative seats reserved for them[13].

This is one of the factors responsible for the lopsided nature of Nigeria's federalism, which saw the North being larger than the three other regions combined. Second, the British eventually handed power at independence to the North which their more educated (at least in western sense) counterparts from the south saw as unjust. Also some saw the Northern groups as 'invaders' because they had conquered the indigenous people. The North was also accused of not having any requisite resources to contribute to the federal pool[14]. For instance in 1912, Southern Nigeria had a revenue of £2.25m and a surplus of £1m while the north had half a million pounds internally generated revenue. This lopsided federalism also partly accounted for the preponderance of Northerners in the armed forces. All these have grave implications for the enterprise of nation building and the conflicts in the Niger Delta as a whole.

Nigeria came into being on January 1, 1914 when Lord Frederick Lugard amalgamated the Northern and Southern protectorates[15]. Before this time what is now known as Nigeria was made up of disparate and distinct ethnic groups. Some of these ethnic groups were organized as Kingdoms.

The amalgamation did not mean the unification of Nigeria. Before the amalgamation, the various ethnic groups in Nigeria pursued separate development patterns. For instance, the North was under the influence of a Muslim reformer Usman Dan Fodio who conquered a large proportion of the northern region in 1804. Apart from this, the Northern part of Nigeria had early contact with North Africa, the Middle East and the Orient in general. The North had also established monarchies, with centralized forms of administration.

They had well-established administrative machinery and were well on their way to conquering and converting the South. They were well versed in Islamic education and jurisprudence. The meaning of all these is that the North had a culture, religion, and politics that were different from that of the South. At this time they had little or no contact with the Europeans and with southern Nigeria as a whole. To make matters worse the colonialists were not interested in building a State but a Colony. Hear Hugh Clifford, a colonial governor of Nigeria:

> Assuming...that the impossible were feasible – that this collection of self-contained and mutually independent Native States, separated from one another, as many of them are, by great distances, by differences of history and traditions, and by ethnological, racial, tribal, political, social and religious barriers, were indeed capable of being welded into a single homogeneous nation – a deadly blow would thereby be struck at the very root of national self-government in Nigeria which secures to each separate people the right to maintain its identity, its individuality and its nationality, its government; and the peculiar political and social institutions which have been evolved for it by the wisdom and by the accumulated experience of generation of its forebears.[16]

In spite of all these, the north was not monolithic. Some ethnic groups in the north have resisted the onslaught of the Muslims. These ethnic groups are the minorities found in the north today. They have remained largely Christians. They include the Tiv, Jukun, Birom, Angas, Bachama, Idoma, Nupe and so many others. They have been at the receiving end of the various religious uprisings in Nigeria[17].

The south had contact with the Europeans in 1444, through a Portuguese adventurer named Lancarote de Freitas. This contact led to the slave trade. And when the slave trade was abolished in 1807[18], the trade in commodities, especially palm oil and palm kernel commenced. In order to control trade, the British needed some form of administrative control. This later led to colonialism.

The south was also exposed to Christian missionaries' activities from Europe. From the missionaries they had western education. This is not to suggest that the south was monolithic. The south had its own share of minorities. But generally, the south was predominantly Christian. The influence of the Europeans depended on several factors, the most crucial being proximity to the coast and availability of raw materials. If an ethnic group was located close to the Atlantic coast, then it was bound to have more contact with the Europeans. If it also had raw materials, which the colonialists needed, then there will be contact. On the other hand if a community was located in the hinterland or had no raw materials, contact was limited.

The south can be divided into two major groups. First is the southwest, which is made up predominantly of Yorubas. Other ethnic groups in the southwest include some that are in the Niger Delta today such as the Ilaje, Ijaw, Itsekiri, Urhobo, Edo, Ukwani and many others. The second group is the southeast, which is made up of mainly Igbos. The southeast also had minority groups such as the Ijaw, Andoni, Ogoni, Ikwerre, Efik, Oron, Ibibio and so many others.

The oil wealth that drives Nigeria is gotten from the minority ethnic groups of the southeast and southwest, while the agricultural products which the north is reputed for, is gotten from mainly the minority areas of the north. The paradox of the Nigerian situation is that the oil wealth, which the majority ethnic groups control and enjoy, is produced in the minority areas. They also bear the brunt of the hazards of producing the wealth. This has been at the root of the agitation of the people of the Niger Delta. The reason for this is obvious. Because they are the minority, they cannot muster the requisite number to influence policies both in a military or democratic regime[19]. This may be one of the reasons why the people of the ND seem to have very little regard for democratic structures.

Before the amalgamation in 1914, the Royal Niger Company through its founder George Goldie Taubman had administered Nigeria under a charter of the British government. And in order to control trade and maximize profit, he wielded the various ethnic groups together through forced and false treaties, conquest and subterfuge. With his newfound wealth, he began to move up north. As he conquered, he annexed. His conquest did not satisfy him; he decided to cut off people of the Niger Delta from their middleman position in the trade with Europe. The people of the Niger Delta did not let this go. They reacted in several ways. First, they tried letter writing, it did not work, and they tried establishing Courts of Equity[20] to regulate trade that did not work either. They sought protection under the British through treaties, they were betrayed. Frustrated, they attacked the trading posts along the coast in order to retrieve any resemblance of dignified livelihood to which they were entitled.

In response, the British massacred about 2000 men, women and children in 1895 at Nembe. The rebellion was perhaps one of the reasons why the British set up the West African Frontier Force in 1898.

The British government considered their new territory too valuable to be left for a company to administer. They revoked the charter of the Royal Niger Company in 1899. While this was going on in the southeast and north, in the southwest Gilbert Carter was also conquering and annexing the various Yoruba Kingdoms. On January I, 1900 the British had three protectorates in Nigeria namely: the Northern Protectorate, Southern Protectorate and the Colony of Lagos. These three were amalgamated in 1914 to form modern Nigeria.

Picture what modern Nigeria is. The north was a theocratic state based on Islam. The southwest was made up of Kingdoms with centralized monarchies that ruled through a combination of checks and balances and the highhandedness and charisma of the reigning King. The east was made up of small clans, villages and communities that were largely republican and egalitarian, while Lagos was cosmopolitan with emerging elite with European tastes and lifestyle. More to this was that Lagos had become home to free slaves from Sierra Leone, Liberia and parts of Europe and North America. This was what formed the modern state of Nigeria.

As if to foretell what was in store for the people of the Niger Delta, in the same 1914 Lugard passed the Colonial Mineral Ordinance. This ordinance made the mining of oil and other minerals in the colony an exclusive preserve of the British. In 1937 another ordinance was passed which gave Shell the monopoly of exploring oil in Nigeria. And in 1938 Shell was granted license to prospect for oil all over Nigeria. A year after Shell started exploiting oil from Oloibiri; the British enacted the Petroleum Profits tax, which was to allow the Nigeria government and Oil Company (Shell) share oil revenue on fifty-fifty basis[21]. The 1960 independence constitution also provided for a Niger Delta DevDevelopment Board (NDDB). This was aimed at addressing the disappointment of the minorities who felt betrayed by the report of the Willink Commission.

Nigeria was granted independence in 1960[22]. At independence, Nigeria had three regions namely North, West and East. And each one of these regions had a political party that dominated the area. And the majority ethnic groups in the regions controlled these dominant parties. The north had the Northern People's Congress (NPC) founded in 1949, which was led by Ahmadu Bello, the grandson of the Muslim reformer Usman Dan Fodio; the West had the Action Group (AG) founded in 1951 and which was led by Obafemi Awolowo, while the East had the National Council for Nigerian Citizens (NCNC) founded in 1944, and led by Nnamdi Azikiwe.

Apart from the NCNC, none of these parties made any pretensions to being national both in outlook and orientation. In 1951, Nnamdi Azikiwe won the elections in Lagos[23] but on the eve of the inauguration of the new government, Obafemi Awolowo allegedly engineered the infamous carpet crossing, which finally nailed Nigeria to the bondage of ethnicity. With this move the NCNC retreated to the east where it dominated until the military struck in 1966.

In each of these regions the minorities also had their own 'small' political parties. In the North there was the United Middle Belt Congress (UMBC) led by Joseph Tarka. The north also had the Northern Elements Progressive Union (NEPU) led by Aminu Kano. One of the unique things about NEPU was that it was founded by someone who was from the majority ethnic group in the north. The difference was in its left-wing ideological orientation. The west had also the Midwest Democratic

Front. The east had other smaller parties too. Within the various regions, the minorities were often marginalized.

Of course the minorities did not accept this state of affairs. They continued to agitate for regions of their own. For instance, just before independence in 1960, the government set up the Willink Commission in 1957 to look into the fears of minorities. In its report, the commission concurred that the fears of minorities were genuine but that it was not enough to warrant the further creation of more regions in the country.

However, the commission recommended that constitutional guarantees be put in place to protect the rights of minorities and allay their fears. This led to further protests by the minorities[24]. These protests were either put down or sabotaged. However, politicians found another use for the minorities' protest. They decided to deal with each other by supporting minorities' agitation in regions other than theirs.

In 1963, the Government of Abubakar Tafawa Balewa created the Midwest Region from the Western Region ostensibly to grant the minorities in the region their self-determination, but the real motive was to politically weaken the Western Region, and undermine the influence of Chief Obafemi Awolowo, who was the opposition leader. This was in retaliation for AG's alleged support for, and financing of minority movements in the east and north while discouraging the same in its power base – the west.

With the 1966 elections in mind, the NPC wanted to improve its chances of being able to form the government at the centre without going into alliance with any other party. The NPC made forays into the various regions. The most hardly hit was the West where the NPC found a ready and willing ally – Akintola, Awolowo's deputy. The NPC was said to have engineered a rebellion within the AG and massively rigged the elections in the region. People took to the streets to protest their stolen mandate. In the ensuing confusion the military struck on January 15, 1966. At the end of the day, Akintola was killed; Tafawa Balewa and the leader of the NPC and premier of Northern region Ahmadu were also assassinated. Many others were killed.

But the coup was not successful. The remnants of the civilian regime hurriedly handed over power to the military hierarchy. Aguiyi Ironsi, an Igbo emerged the new Head of State. The coup was interpreted as having an Igbo flavour since almost no Igbo politician was killed. Second, many of the coup leaders were Igbos. In retaliation, northern officers struck in July 1967. One of the causalities was Aguiyi Ironsi, the new military Head of State.

Yakubu Gowon, a thirty-two year old officer from the northern minority Angas ethnic group was appointed Head of State. Ojukwu, the Oxford-educated Military Governor of the Eastern region refused to recognize him. This triggered a chain of events that eventually led to the declaration of a Republic of Biafra by the East and a 30-month civil war, from 1967 to 1970

In 1970, the civil war ended. The Biafrans were defeated and brought back into the Nigerian federation.

In December 1983, Mohammadu Buhari became the new head of state after overthrowing the civilian regime of Shehu Shagari. The Buhari regime was in turn overthrown after barely a year in office by the regime's Chief of Army Staff – Ibrahim Babaginda (IBB). Two significant events that affected the conflicts in the Niger Delta took place under IBB.

First, in 1990 the people of Umuechem,[25] an oil-bearing community in eastern part of the Niger Delta complained of neglect by the oil company operating in their area. In the ensuing confusion, the police marched on the community. At the end of the day 80 people, including the traditional ruler of the community and a

Methodist minister lay dead. More than 400 houses were destroyed.

Second, IBB established the Oil Minerals Producing Areas Development Commission (OMPADEC) in 1992. This commission was to address the development needs of the Niger Delta. Its success and failure shall be discussed later. Suffice to mention that it did not outlast the regime that established it.

In 1993 an election was held which saw the multimillionaire businessman Abiola coasting home to victory. IBB annulled the elections in a terse and unsigned statement. That singular act threw Nigeria into a crisis. In the ensuing confusion, IBB appointed an interim government and stepped aside. Ernest Shonekan, a former director of Shell and former Chairman of the United African Company (UAC) – grandchild of the Royal Niger Company, headed the interim government. Shonekan and his interim contraption did not last long enough. In November 1993, Abacha, the Chairman, Joint Chief of Staff and Minister of Defence overthrew the interim government.

It was during Abacha's dictatorship that Nigeria witnessed the height of personalization of governance. It was during Abacha's regime that Ken Saro Wiwa and eight others were executed. The dynamics of conflicts in the Niger Delta was redefined and made part and parcel of the democratization process. However, Abacha established the Petroleum Trust Fund (PTF) in 1994. This was an extra-ministerial body that was to use the savings from the removal of petroleum subsidy as decreed by the World Bank and IMF to develop the Niger Delta. It is instructive that it was established in response to the protests over the increase in the pump price of petroleum products and not for the special care of the Niger Delta. At the end of the day, PTF may have carried out more development projects in non-oil bearing regions than in the Niger Delta.

In 1998 Abacha died on his bed. Abdusalami Abubakar, another General, took over from him. Abubakar started the transition that eventually led to the inauguration of the present civilian administration headed by former military head of state, Obasanjo.

Under Obasanjo two particularly important incidents have happened that affected the Niger Delta. First, was the destruction of Odi in November 1999 by a detachment of the Nigerian military, ostensibly in search of the killers of twelve policemen by a gang of youths. Odi is located in the oil-bearing state of Bayelsa in the Niger Delta. This community of about 15,000 was razed[26].

The second was the establishment of the Niger Delta Development Commission (NDDC) in 1999. This body is to address the development needs of the Niger Delta. Since inception, it has been mired in controversy[27]. First, was which states of the federation would be considered Niger Delta states? Second, was the issue of who will contribute what? Third was who shall head the organization. But the most important was that the people of the Niger Delta, especially the politicians, opposed its establishment. They argued that the money should be given to them to develop their area instead of creating a new parastatal.

The Niger Delta: An Overview

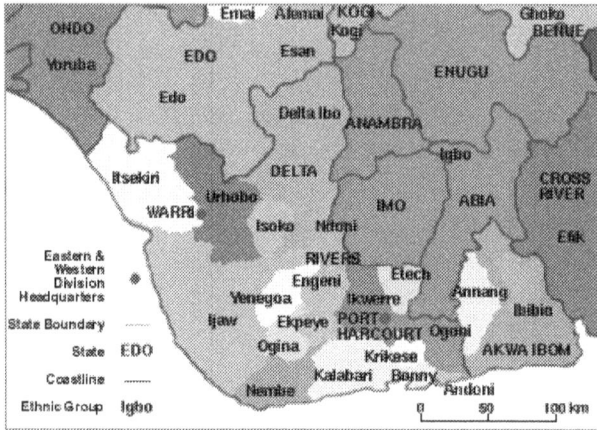

Map of the Niger Delta showing the different ethnic groups: Courtesy of Shell.

There is no place known as the Niger Delta (ND)[28]. The ND is simply an agglomeration of several groups that inhabit a contiguous stretch of territory. In this study, the ND refers to the area where oil is explored in Nigeria. The ND is made up of nine states namely: Abia, Akwa Ibom, Bayelsa, Cross River, Delta, Edo, Imo, Rivers and Ondo. However, the volume of oil produced from each state differs. The ethnic groups inhabiting each state are also different. And in Nigeria's peculiar political parlance, there is a dichotomy of Core ND and peripheral states. This to a large extent has more to do with the quantity of oil produced. But some scholars[29] also use ethnic criteria. For instance, some people do not recognize the Igbos as being part of the ND.

The ND is about 70,000 square miles. It is one of the largest wetlands in the world. The area has three main ecological zones namely: the sandy coastal area, the fresh water swamps and the dry forest land areas. There

74

are more than 20 different ethnic groups in the ND. The Ijaw is reputed to be the largest ethnic group in the ND and the fourth largest ethnic group in Nigeria.

Space may not permit an elaborate discussion on the origin[30] of people of the ND. Suffice to say that like most Nigerian ethnic groups they have myths of origin that could either be traced from heaven, the east or the ocean. For instance, Ndoki people claim that they originated from the Atlantic Ocean[31].

The people of ND belong to what has been referred to as high context cultures. In other words they are more communal in lifestyle than individualist. The family is the basic unit of social organization. The father is the head of the family while polygamy is not frowned at. The main occupation is farming and fishing. They produce mainly tubers like yam, cocoyam and cassava. They also grow palm trees and other fruits like pears, mango, paw-paw etc.

Some ethnic groups in the ND have centralized political organization while others are organized in small community units. For instance, The Itsekiris of present day Delta state have an elaborate kingship system, while the Ndoki may be classified as "stateless society"[32].

The age set system is highly developed among people of the ND. Elders play a very great role in the communities of the ND. Age is respected and revered. The people are mainly traditional worshippers. They have four different categories of the world.[33] First, they have the world of the unborn, second the world of ancestors, third, the living dead and fourth, the living.

They believe that for order, harmony, peace and progress, the relationship between these four must be cordial. But with the advent of Christianity many have converted but still kept some of their traditional beliefs and practices like fishing festival and the almost mystical relationship with the rivers.

To an average ND indigene conflict is bad and should be avoided[34]. Because they belong to the high context group of cultures, group harmony and solidarity is very much valued. They have an elaborate and intricate web of indigenous conflict resolution mechanisms[35]. Conflicts or disagreements are usually handled at the family level. If they are not satisfactorily resolved, they are handed over to either the daughters of the compound or the married women. If all these fail, the elders take over. Still if resolution is not successful the matter is referred to the entire community gathering of adults.

Sanctions for non-compliance with societal norms could be very severe[36]. They range from ostracisation, to banishment, exile, slavery and fines. The gods and ancestors played a great role in community issues. There is also an elaborate ritual of reconciliation.

It is perhaps necessary to point out that a community that is so close to nature, and where every individual counts, is at the moment more mythical than real. People remember it with a forlorn sense of nostalgia. In its place is a community that is struggling to define itself and make a sense out of its present predicament.

No one can say exactly when the change began or when it is going to end, if it ever will. No one is sure of the main causes of the change. But there is a consensus that the ND is no longer what it used to be, and will never be the same. Fingers point to as far back as 1444 when the first Portuguese Freitas arrived the area and in return for the hospitability of the people, carted away 235 people who he sold in Europe as slaves. That began the obnoxious evil called the slave trade. Right on the heels of this was Christianity and western education. Before they could catch their breath colonialism landed. All these were to fundamentally alter the structure of the ND communities.

It has been argued that conflict is a child of history. One is inclined to say that conflict is a product of interaction. Before the coming of the white man the communities of the ND were interacting. They intermarried, traded with each other, fought against each other and also celebrated with each other. An individual's worth in the ND communities was determined by industry. It was a land of opportunities where even slaves aspired to, and rose to kingship[37].

Change did not only affect the lifestyle of the people, it also affected their perceptions of conflict and peace. To have peace they fought wars. They went to war to capture slaves, wars to prevent themselves from being captured and sold into slavery. They fought to be part of the so-called "legitimate trade". They fought to preserve their communities from being run over. In short they fought for their lives.

Eight different conflicts can be identified in the ND today. These conflicts are neither special nor unique to the ND. However, the intensity and dynamics of these conflicts have been affected both by the presence of, and the politics of oil. The interest in these conflicts is to see whether a theoretical pattern for understanding and analyzing them could be established. Second, is also to know where and how to intervene.

As stated earlier the people of the ND are mainly farmers and fishermen/women. Oil on the other hand is explored both on land and from the sea. With oil exploration and farming competing, land will be in short supply. Like most rural communities whose life is tied to the land, there are many land disputes in the ND.

These land disputes are between families, villages, ethnic groups, governments and oil companies. One of the most publicized land disputes in the ND is between the Eleme and Okrika in Rivers State. The Ijaw, Urhobo and Itsekiri crisis in Delta State is also linked to land.[38]

The cause of these land disputes is not far fetched. Being in the riverine area with a growing population and urbanization, land was definitely going to be in short supply. The influx of oil company workers and their contractors complicated this. Conflicts about land are escalated by the presence of oil on such lands.

More importantly, land has a symbolic significance for the people of the ND[39]. Land is not just a mere resource to be exploited. Land is bequeathed from generation to generation. They bury their dead on the land. They even worship the earth and river goddesses/gods. When settling disputes people bare their feet and have direct contact with the land as a symbol of honesty. They bury the umbilical cord of newly born babies in the land as a mark of identity. They have a psychological attachment to the land.

Today, in the ND, this sacred resource has been desecrated. Land has been demystified and rendered ordinary. According to the people, the abominations being experienced today may not be unconnected with the desecrations of the land[40]. The commercialization of land is something that the people are finding very difficult to come to terms with. For instance a recurring question is how can you sell land?

The Federal Government of Nigeria complicated this land issue when in 1978 they promulgated the Land Use Act. This law vested control and ownership of every piece of land on the government. The import of this is that the people of the ND are now tenants on a piece of land that they inherited from their ancestors. Earlier in 1969, the same government has also promulgated the Minerals Decree, which gave the Federal Government control of oil revenues.

To put all these in context it is important to remember that all these were happening when oil has become the main issue in Nigerian politics. Moreover when the other regions had their palm oil, cocoa and groundnut the regions controlled these resources and used the revenue to develop their regions. The question the people of ND are asking is why us, why now? We shall see later how the formulation of all these impacted their perception of the conflict and how it became a motivating factor for their responses to the conflicts.

It is also important to bear in mind that with the presence of strangers who do not share their beliefs and culture and a distant, esoteric and an anonymous entity (the government) calling the shots, the whole dynamic of land dispute was bound to change. In fact it did change in a very fundamental way with very serious implications for conflict resolution. With strangers on the land and the land belonging to the federal government by implication, the ND person has simply ceased to exist. Even where the ND person decides to do something, they are so disempowered because their land is no longer available for farming, the water is polluted and there is no fish, while the other party in conflict works for a multinational, earning good money from the inheritance of the ND people, with the backing of a government that claims to represent and protect the people of the ND

Another kind of conflict that was identified is chieftaincy disputes. Before colonialism some groups in the ND had Chiefs or Kings as the case may be. The colonialists introduced indirect rule, which meant that those who did not have Kings had one appointed for them, often from the rank of the never-do-wells that hobnobbed with the colonizers. When the time came for Chiefs to be appointed, these were the people that the white man appointed.

Those who already had Kings had a different kind of problem. Some of the existing monarchs were quickly deposed and exiled. Their replacements were often people loyal to the colonial order. This was the situation when the government carried out the Local Government reform of 1976 by Ibrahim Dasuki. The reform created Kingships where none existed and for existing ones, they were to be given Staff of Office (euphemism for recognition) by the government. The local government reform also granted state governments the powers to create autonomous communities – that is, a sort of new kingdoms, with new kings. The Chieftaincy institution therefore became an all-comer's affair. All these were happening against the backdrop of the increasing influence of oil. Government now owns and controls the land, government has appropriated the mineral resources for its sole control, and government has created new kingdoms and appointed kings.

Rothman[41] has argued that when essential group identities are threatened or frustrated, conflict arises. It took the people of the ND time to realize this threat. And Schulz[42] has admonished that social conflict emerges and develops because of the meaning and interpretation people attach to actions and events. Since most of these were happening under military rule and there was no democratic space to access the government, the Chieftaincy institution became the end-it and be-it all for anyone willing to raise his head. And even if democracy was in vogue, the ND did not possess the requisite number to make any appreciable impact. Even if they could make impact they did not have the resources to contest elective offices.

Underlying all these is the fact that the traditional norms have given way and can no longer regulate the conduct and contest for chieftaincy positions. So succession disputes are very prevalent. Only the government can arbitrate this contest. So whoever can manoeuvre his way to the corridors of power even if he is a miscreant could be made a chief. The contest became a cutthroat do or die affair. The dispute becomes intense if the community produced oil. This is because the traditional ruler shall be in a position to interface with the oil companies for such patronages as contracts, jobs and other financial payoffs. A very good example of this chieftaincy confusion is the tragic drama that is playing itself out in Kula Kingdom in Rivers State[43]. Comments on it shall be limited because the matter is still in court. This case started as far back as 1982.

At the beginning of this chapter reference was made to the existence of so many parties in the ND conflicts. Before the conflicts took on a life of its own, the ND society was one that was organized around sets and associations[44]. There were the various age sets, the fishermen guild, the hunters' club and various titleholders. These associations were meant to serve the community, maintain social control and enforce rules and regulations. Leadership in these organizations evolved. It was not by contest. Members who lead were chosen for their special skills and specific talents that were put to the use of the entire society. For instance, to lead the hunters' guild, one must be a good marksman and a brave hunter. Evidence shall be a thing like the head of animals killed in hunting expeditions, crocodile skin and even titles for recognition. These criteria, which were in-built and self-regulating, were to give way to electioneering. This bred opposition camps, which in turn divided the community.

From the foregoing discussion some conclusions could be drawn regarding the conflicts in the ND. First and foremost is that the conflicts in the ND fall into the category of what Edward Azar defines as "protracted social conflict (PSC)"[45]. According to Azar, PSC is "the prolonged and often violent struggle by communal groups [religious, ethnic, racial, or cultural] for such basic needs as security, recognition and acceptance, fair access to political institutions, and economic participation."[46] Azar listed ten propositions of PSC. By PSC he means that such conflicts may never come to a resolution. This is because "the real sources of the conflicts are deep rooted in the lives and ontological being of those concerned". Very important, Azar avers that "those involved in PSC seem to have difficulty in articulating what it is that leads them to violent protest and even war".[47]

Among other features Azar states that the uniqueness of PSC is because of local circumstances, histories and attitudes which give these conflicts their individuality. Azar also emphasizes the need to understand thoroughly the identity of the parties involved in the conflicts. And finally Azar posits that PSC "appear to start with one set of stated goals, primary actors and tactics, but very quickly acquire new sub-actors, new goals and new types of resources and behaviours."[48]

All of the above fits the profile of the conflicts in the ND as was discussed in the chapter. The lesson from it is that the conflicts are man-made, reinforced by unjust social and political structures and embedded in relationships within the system. Conflicts are never independent of the structures of their context. For instance, Ikelegbe[49] argued that the proliferation of civil societies in Nigeria in the 1980s led to the proliferation of organizations in the ND. More to that is that the ND society was one that had, and made extensive use of associations and groups. The conflicts in the ND involve issues that are mutually incompatible and mutually exclusive. For instance, the ND person is finding it difficult to be a Nigerian and at the same time being a ND indigene. On the other hand, their resources could not be used for the entire Nigerian state and at the same time be used exclusively for them.

Second there are values that have been ascribed to the issues involved in the conflicts in the ND. For instance, is oil production more important than fishing and farming? Who decides which activity is more important than the other?

Third, information as it relates to the conflicts in the ND has made some issues latent and others manifest. This has led to an obscuring of several issues by third parties and fronts. For instance, almost every Nigerian politician, leader or citizen agrees that the people of the ND are not being fairly treated, but they never want to do anything about it.

Peter Coleman[50] identified four main causes of intractable social conflicts. They are issues, context, dynamics, complexity, social-psychological and parties. He argued that the changing pattern of all of the above causes a conflict to become intractable. All these we have seen in our analysis of the conflicts in the ND. Morton Deutsch[51] takes the argument further by stating that internal conflicts actually perpetuate external ones. What this means is that while ND people are in conflict over land, chieftaincy and leadership, this perpetuates their conflicts with the oil companies and the Nigerian state.

Finally, a critical issue that arises is the over-politicization of the discontent in the ND. As issues get increasingly politicized, the conflict assumes a more intense dimension, the interests of the parties become more confounding and obscure and their positions unclear. This is what Augsburger[52] refers to as the uncontrolled escalation of conflict, which is self-perpetuating and cyclical in nature. This has made it almost impossible to intervene in the conflict as a neutral third party. Clearing all of the above issues, and addressing the conflicts, bit-by-bit, community-by-community, issue by issue, and from time to time, seems to be a more feasible approach. The next challenge is to demystify the conflicts from the charge of obscurity and impenetrable density. The next chapter will address that.

Conclusion

From the foregoing some insights could be drawn to illuminate this study. First, oil is one of the main issues, if not the main issue, in Nigerian politics. Every government from the colonialists to the politicians to the military all paid attention to the issue of oil. Why is it that in spite of all their attention, the bird is still crying?

Second every government in Nigeria has passed one form of legislation or established one body or the other to deal with the discontent in the Niger Delta. Again in spite of all these, instead of a reduction in the protests, it is still escalating. Third, very little of the initiatives so far came from the people of the Niger Delta. They all have been imposed from outside. The initiatives tend to give one the impression that the people of the Niger Delta do not know what they want. Meanwhile, they have shouted themselves hoarse and no one is listening.

Finally no attempt has been made to resolve and transform the conflicts in the Niger Delta. All we see have been "development dumping".

Notes and References

[1] Quoted by C.K.Meek in "The Niger and the Classics: The History of a Name. Journal of African History, Vol. 1, No. 1, 1960, p.1

[2] Falola, Toyin. The History of Nigeria. Westport, Connecticut: Greenwood Press, 1999, pp. 1-4.

[3]Gary, Ian. Bottom of the Barrel. Baltimore, Maryland: Catholic Relief Services, 2003, p.27.

[4]See Mark Anikpo, "Social Structure and the National Question in Nigeria" in Momoh, A. and Said Adejumobi (Ed.) The National Question in Nigeria: Comparative Perspectives. Burlington, Vermont: Ashgate Publishing, 2002, p.51.

[5]See Chinweizu. The West and the Rest of Us. New York, Random House: 1978. See also Rodney, Walter. How Europe Underdeveloped Africa. London, Bogle-L'Ouverture Publications: 1981.

[6] See Harrison, L.E. and Samuel P. Huntington (Ed.). Culture Matters: How Values Shape Human Progress. New York: Basic Books, 2000.

[7] See the documentary "The Triple Heritage" by Ali Mazrui produced by the BBC.

[8] See Boiling Point, p.75.

[9] See Frynas, Jedrzej Gerorg. Oil in Nigeria: Conflict and Litigation Between Oil Companies and Village Communities. Hamburg: LIT, 2000. P.66

[10]See Crawford Young "The Heritage of Colonialism" in Harbeson, J.W. and Donald Rothchild (Ed.) Africa in World Politics: The African State System in Flux. Boulder, Colorado: Westview Press, 2000, p.25.

[11] See Nwabueze, B.O. A Constitutional History of Nigeria Harlow: Longman Publishers: 1982, P.127.

[12] See Osaghae, E.E. Crippled Giant: Nigeria Since Independence. Bloomington, Indiana: Indiana University Press, 1998, pp. 6 & 7.

[13] See Tamuno, Tekena N. "Separatist Agitations in Nigeria Since 1914". The Journal of Modern African Studies, 8, 4 (1970) pp.563-84.

[14] See Perham, Margery. Lugard: The Years of Authority (1898-1945). London: Commonwealth Publishers, 1960, p.418.

[15] See Ballard, J.A. "Administrative Origins of Nigeria's Federalism", Journal of African Affairs, Vol. 70, No.281, 1971, p. 334.

[16] Governor Hugh Clifford's address to the Nigerian Council on December 29, 1920. Document is available at National Library Lagos, National Archives Enugu and Ibadan.

[17] See Kukah, M.H. Religion, Politics and Power in Northern Nigeria. Ibadan, Nigeria: Spectrum Books, 1994.

[18] See Davidson, Basil. The African Slave Trade. Boston: Cambridge University Press: 1961, pp. 252-253.

[19] See Osaghae, Eghosa E. "The Ogoni Uprising: Oil Politics, Minority Agitation and the Future of the Nigerian State." African Affairs, 94, (1995) pp.325-344.

[20] See Ikime, Obaro. The Fall of Nigeria: The British Conquest. London: Heinemann, 1982.

[21] Okonta, Ike and Oronto Douglas. Where Vultures Feast: Shell, Human Rights and Oil in the Niger Delta. San Francisco, CA: Sierra Club Books, 2001.

[22] See Okadigbo, Chuba. Power and Leadership in Nigeria. Enugu: Fourth Dimension Publishers, 1987.

[23] Dudley, B.J. "A Coalition Theoretic Analysis of Nigerian Politics (1950-66), The African Review, Vol.2, No.4, 1974, p. 532.

[24] For a detailed analysis of some of these protests see Anifowose, Remi. Violence and Politics in Nigeria: The Tiv and Yoruba Experience. New York: Nok Publishers, 1982.

[25] See Browne, M. The Price of Oil: Corporate Responsibility and Human Rights Violations in Nigeria's Oil Producing communities. New York: Human Rights Watch, 1999.

[26] For details of the Odi incident see Human Rights Watch. "The Niger Delta: No Democratic Dividend", Human Rights Watch Country Report, Vol. 14, No.7 (A), October 2002, p. 21.

[27] See Thisday Newspaper, Tuesday, August 6, 2002, p.1

[28] See The Guardian, September 1, 1999, p.56.

[29] See Osaghae, E.E. Crippled Giant: Nigeria Since Independence. Bloomington, Indiana: Indiana University Press, 1998.

[30] For a detailed study of the origin of the various ethnic groups of the ND see Alagoa, E.J. A History of the Niger Delta:

An Historical Interpretation of Ijo Oral Tradition. Heinemann: London, 1972.

[31] Interview with a Chief from the area in June 2003 for the CDA STEPS project.

[32] Evans-Pritchard used this to describe societies that do not have centralized political organization. See Fortes, M. and E.E. Evans-Pritchard. African Political Systems. London: Oxford University Press, 1940.

[33] Interview with an old woman of about 90 year simply called Npa from Asa.

[34] Interview with Lady Victoria Obinya, wife of the Anglican Bishop of Ukwa and President Mother's Union.

[35] Interview with His Royal Highness Eze Innocent Nkwocha (a traditional ruler in the ND).

[36] Interview with Chief G.I. Akara, a community leader of Ndoki in Ukwa East Local Government of Abia State.

[37] King Jaja of Opobo is a good example. See Ikime, O. The Fall of Nigeria: The British Conquest. London: Heinemann Books, 1982.

[38] See Imobighe, T.A. et al. Conflict and Instability in the Niger Delta: The Warri Case. Ibadan: Spectrum Books, 2002.

[39] Interview with Mrs Julian George Joseph from Ndoki.

[40] Interview with His Royal Highness, Eze Nkwocha of Asa.

[41] Rothman, J. Resolving Identity-Based Conflict in nations, Organizations and Communities. Jossey-Bass: San Francisco, 1997.

[42] Schulz, A. The Phenomenology of the Social World. Northwestern University Press, Evanston, Illinois, 1967.

[43] Interview with a young man released from police detention over the crisis. He does not want his name in print.

[44] Interview with Chief Don Ubani of Asa.

[45] Burton J. and Frank Dukes. Conflict: Readings in Management and Resolution. New York: St. Martin's Press, 1990, p.145.

[46] Azar, E.E., "The Analysis and Management of Protracted Conflicts", in The Psychodynamics of International Relationships, Vol. 2, ed. Vamik D. Volkan, Joseph V. Montville, and Demetrios A. Julius. Lexington, Mass. Lexington, 1991, p.93.

[47] Ibid. p.146.

[48] Ibid: p.154.

[49] Ikelegbe, A. Civil Society, Oil and Conflict in the Niger Delta Region of Nigeria. Pp.437-469.

[50] Peter Coleman. "Intractable Conflict", in Morton Deutsch and Peter Coleman, eds. Handbook of Conflict Resolution (San Francisco: Jossey Bass), 2000, p.432

[51] Ibid: 432.

[52]Augsburger, D.W. Conflict Mediation Across Culturs: Pathways and Pattern. Westminster/John Knox Press: Louisville, Kentucky, 1992.

Chapter Three

Review of literature on the Conflicts in the Niger Delta

The Niger Delta (ND) of Nigeria is an over-researched area'. There are many books and articles explaining the conflicts in the Niger Delta. The main reason for this is the presence of oil in the area. Second, there has been an increasing interest by the international community in environmental and human rights issues. And all these could be found in abundance in the conflicts in the ND.

The consensus among scholars[1] however is that the ND deserves a better deal from the Nigerian nation. The point of disagreement is what, why and how? The first step towards the resolution and transformation of a conflict is to properly and clearly explain it. This explanation has been lacking in the study of the ND. One possible reason is probably because conflict resolution is a relatively new discipline especially in Nigeria. Another possible reason is that people have not seen what is happening in the ND as conflict. It has been seen either as part of the political evolution of the Nigerian nation or part of the history of the ND. This perhaps also explains why most solutions seem not to be working. This chapter will examine some of these explanations and solutions and the justifications behind them.

There are three main schools of thought on the conflicts in the ND. First are those who argue and believe that the oil companies represent the worst form of evil and that the governments of Nigeria encourage the oil companies to behave irresponsibly. In this school are such activists as Oronto Douglas, Ike Okonta, Odia Ofeimun and others. As we go on in this study, their views shall be espoused.

The second group is those that argue that the people of the ND are unreasonable and insatiable. This view is usually muted since it will be politically incorrect to go public with this kind of viewpoint especially after a visit to the ND. Those who espouse these views are mostly the oil companies and their staff. Some government officials also hold these views. For instance, the views of Precious Omoku, Shell's Corporate External Relations Manager[2] could fall into this category. People who hold these views do not come out openly to condemn the people of the ND; they come more in the form of defence of oil company activities and operations.

The third group is those who I will refer to as the "conspiracy theory" school. This group believes that there is a grand conspiracy by international capital, the Nigerian elite, dominant ethnic groups and others to finish off the ND[3]. This group submerges the conflicts in the ND into the political evolution of the Nigerian nation. Such issues as corruption, class struggle, failed state, bad governance, bad constitution, improper practice of federalism and many others are lumped together. On the face of it, this school may have some similarities with the eclectic school of conflict. But they are not. This is because the conflicts in the ND could be explained using the above parameters. Professor Julius Ihonvbere, Dr. Dan Omoweh and Dr. Festus Iyayi all could be said to belong to this school. This school also draws a lot from Marxism.

What has been missing is what I will refer to as the conflict transformation school. This is the school that will analyze the conflicts in the ND based on the social interaction process and relationship patterns of the parties in the conflict in the ND with a view to putting in place mechanisms for transforming the conflicts. This school does not engage in the blame game. This school is more interested in building the right kind of relationships between the parties and stakeholders in the ND. This book falls into that category.

The book will build and give a framework to the approach of Rev. Fr. Kevin O'Hara under the auspices of the Centre for Social and Corporate Responsibility (CSCR). The unique feature of this school is that it is following a triangular approach to understanding and transforming the conflicts in the ND. This school's approach shall be examined in detail in the subsequent chapters. But meanwhile let us discuss how some of the schools have conceptualized the conflicts in the ND. The aim here is to analyze some of these views in order to see how they fit into the over all picture of the conflicts in the ND.

Many views of conflicts in the ND have been driven more by populism than by any academic rigour and practical experience. Moreover, academics from outside of the ND tend to ignore the conflicts there since the conflicts in the ND are not unique. We should bear this in mind as we discuss these views.

Colonialism is the favourite whipping boy of most African and Pan-African scholars[4]. Colonialism has been blamed for the woes of the people of the ND. According to Osarhieme Osadolor the guilt of colonialism did not lie in the creation "of the forces and conditions which tended to fragment the Nigerian nation-state, as it was that of manipulators and exacerbation of those diverse factors which differentiated the country"[5]. Festus Iyayi, a Niger Delta indigene and a professor of business administration, took the argument a step further when he declared, "It needs to be noted however, that fundamentally, the problem of the Niger Delta arises from the character of the neocolonial dependent Nigerian capitalist economy"[6]. The argument is that the people of the ND have accumulated grievances that date back to the pre-colonial and colonial times. In this view therefore, the conflicts in the area today are a manifestation of the unsolved grievances of the past. They argue that the colonialists did not allow the indigenes of the ND to participate in the slave trade. They also denied them their position of middlemen during the period of 'legitimate' trade[7].

Moreover with the amalgamation of northern and southern Nigeria in 1914, the major ethnic groups dominated the ND. To buttress their argument they point to the relic of the colonial past - the Oil Companies as represented in the ND today. This argument is not without justification. The root of most conflicts within the modern state structure in Africa today could be located in and around colonialism. According to Crawford Young, "A final legacy of the colonial system is the series of regional crises it has left in its wake...."[8]Its progenitor neo-colonialism is still on the prowl. Evidences abound.

But the danger in this mindset is that it relieves Africans of the responsibility of confronting the issues affecting them. And in the ND it provides an escape route for Nigerians to blame someone else. And for conflict resolution specialists and scholars, the colonialism argument disenables. This is because it makes the conflict seem overwhelming. Second, while colonialism may provide a valid political and economic explanation for what is happening in the ND, it may be limited in the area of conflict analysis and resolution.

This is because it does not draw attention to the relational issues involved in oil exploration activities. Moreover, conflict resolution interventions do not start from the root causes since those are systemic issues. This is because even if the systemic issues are addressed it still leaves the relational issues unattended.

Colonialism has happened and there is nothing anyone can do about it. It is a fact of history and reality of African existence. Neocolonialism is also real because it limits the potentials of former colonies to take initiatives that will better their lives. But Africa (nay ND) cannot afford to remain a prisoner of her past. Many countries in Asia have been able to break away from the colonial logjam. The Igbos say that if a child washes his hands clean that child will dine with elders. ND does not only need to wash her hands clean, but to wash colonialism off it. According to Robert Dibie a public policy specialist, "The real powers behind Nigerian politics since independence have been (a) top political party leaders before 1966; (b) military leaders since 1966; and (c) senior bureaucracy throughout the past four decades"[9]. Time has come to use this power to solve the ND issue.

Moreover, the colonialism argument tends to give the impression that Africans are incapable of influencing a change process from within. This ability to influence a change process from within is what Whitaker[10] refers to as "autonomous capacity." Africans have the autonomous capacity to generate change and sustain it. That is one of the aims of this book.

The problem for now is no longer colonialism but as the cerebral Nigerian novelist Chinua Achebe puts it "The problem with Nigeria is simply and squarely a failure of leadership. The Nigerian problem is the unwillingness or inability of its leaders to rise to the responsibility, to the challenge of personal examples which are the hallmarks of true leadership"[11]. Therefore three critical issues are required to address the conflicts in the ND. First is autonomous capacity, second is leadership and third is the enabling environment or social space for intervention.

Closely related to the colonialism explanation is the issue of federalism. The British introduced Federalism in

Nigeria in 1947 when they created the three regions. Till date Nigeria is still a federation. According to scholars[12], federalism is meant to address the fear of minorities. It is also designed to ensure that no part of the federation is big enough to hold the rest of the federation to ransom[13]. However, because of the various development crises rocking Nigeria, they argue that Nigeria's federalism is not a proper one. Two Nigerians have clearly articulated the ND idea of a proper federation. First, is Obafemi Awolowo who insisted that "ethnic groups and nationalities would form the basis of the federation"[14]. This is in contradistinction to the notion of artificial boundaries of creating states. The second proponent is Ken Saro Wiwa, the slain Ogoni rights activist who referred to Nigeria's federalism as "internal colonialism".[15] According to Odia Ofeimun:

> "Essentially, the Awolowo strategy was in diametrical opposition to a unitary ethic which assumed a nationality that still needed to be properly negotiated, his position called for existing regions to be broken up in favour of ethnic groups"[16].

This proposition is curious since Awolowo never supported the creation of any regions, states or provinces from his heterogeneous west[17]. Second, how many regions would Nigeria have had from more than 250 ethnic groups if the creation of states were based on ethnic nationality? Finally, there is nothing in the proposition that recommends it both as an explanation and solution to the ND. For instance, Ondo State is regarded as being part of the Niger Delta. It is largely homogeneous. However, there are conflicts between the Ilajes and Ijaws and the oil companies in the state[18]. Moreover, a region, state or province that is ethnically homogeneous does not guarantee that they will not have conflict with oil companies or that they will not be impacted by oil operations.

In other words Nigeria's federalism is lopsided. "The problem with the federal structure was that it not only inequitably incorporated minorities into ethnic-dominated regional bastions, but also created a disproportionately large northern region, which included nearly three-quarters of Nigeria's territory and over half its population"[19]. So the agitation of the people of the ND has been centred on the 'restructuring of the Nigeria federation'. And by restructuring they mean the granting of further autonomy to the federating units. They also mean that the constituent units should be allowed to control the resources in their area. But Rothchild and Harbeson counter that one of the main features and in fact functions of a modern state is to extract resources within its jurisdiction and use them for the betterment of the citizens[20]. Nigeria as a nation state has extracted resources from the Niger Delta. But whether it has used it for the betterment of the citizens of the ND is part of the challenge of this discourse.

The argument of federalism as an explanation or cause of the ND issue is plausible if the issue at stake is political engineering. Evidence on the ground shows that federalism have been used to address the conflicts in the ND[21]. Three regions were created in 1947, in 1963 the fourth was added, in 1967 twelve states were created, in 1976 the states were made 18; in 1991 they became 30 and 36 by 1996[22]. As more states were created so was the escalation of the conflicts in the ND. If the creation of more states has not been able to address the issue since almost 60 years ago, it then means that something is lacking in the federalism argument. Uzoigwe argues "the response of the minority groups to Nigeria's ethnic politics was to demand the creation of states. Practically all these demands have been met ...

the minority groups have their own future in their hands as it were...."[23]

There are three broad applications of the principle of federalism to conflict resolution. First, federalism serves as a link to the Central government. In this instance, federalism becomes the lubricant of the relationship between the various tiers of government. Second, federalism is used to unite divided societies while each entity still maintains their independence in non-negotiable issues. The third is the use of federalism to achieve the economy of scale. That is each group or branch produces what it has comparative advantage in.

From the above it is obvious that federalism is neither a cause nor explanation for the conflicts in the ND. Rather federalism has been tried out as solution. Subsequent leaders[24] in Nigeria who tried to build a modern state from disparate and distinct ethnic groups have used Federalism. So we need to look elsewhere for a more plausible explanation.

The federalism argument gained currency under the IBB and Abacha regimes especially after the June 12, 1993 elections were annulled. In 1967 the Igbos went to war for greater autonomy. The ND neither supported the war nor used the opportunity to fight for autonomy. Before the war, Ojukwu had extracted an agreement for greater autonomy for the regions at the Aburi meeting. When Gowon renounced the agreement, the people of the ND did nothing to either support Ojukwu or persuade Gowon to honour the agreement. From the foregoing it is clear that the federalism argument is either an after thought or a fad in line with political sloganeering in Nigeria. For instance, how come that the people of ND have consistently voted for "northern parties" that do not agree with the restructuring agenda?

The conflicts in the ND have also been described as ethnic conflicts[25]. According to Sam G. Amoo "the most popular and enduring perspective on the sources of conflicts in Africa is the contention that ethnicity per se constitutes the critical, if not the determinant source of conflicts on the continent[26]". Mainly those who think or believe that it is impossible for different ethnic groups to co-exist espouse this argument. According to Thomson an ethnic group "is a community of people who have a conviction that they have a common identity and common fate based on issues of origin, kinship, ties, traditions, cultural uniqueness, a shared history and possibly a shared language".[27] From this definition, the ND does not qualify as an ethnic group. Moreover, Edward Azar has argued "it is the denial of human needs, of which ethnic identity is merely one, that finally emerges as the source of conflict, be it domestic, communal, international or inter-state."[28]

Another viewpoint is that ethnicity is a social construction especially used by the elite to mobilize the masses in times of conflict. Brass argues that "they are the creations of the elite, who draw upon, distort and sometimes fabricate materials from the cultures of the groups they wish to represent in order to protect their well-being or existence or to gain political and economic advantage for their groups as well as for themselves".[29]

A critical analysis of the ND conflict will show that it has little or nothing to do with ethnicity. Bostock tells us that ethnic conflict is a "breakdown of accommodation of ethnic minorities within a state".[30] Perhaps the only thing ethnic about the conflicts in the ND is the inter-communal clashes between the various ethnic groups in the ND. The people of ND do not even recognize these inter-communal conflicts as issues. They believe that their main problems lie with the oil companies and governments. Come to think of it, the ND is not even ethnically homogeneous. For instance, the general believe is that Ogoni is one homogeneous ethnic group. According to Osaghae, an academic and indigene of the ND, the "Ogoni is made up of Khana, Gokana and the

Eleme. The Ogoni do not have a myth of common origin, and constitute an ethnic group only on the basis of sharing common language, culture, tradition, farming methods and similar attitudes"[31]. This means that even if autonomy is granted today, a new wave of minorities will be created. For instance in the present Rivers State, the Ogonis accuse the Ijaws of domination.

Closely related to this is the ethnic minority issue. In the first place, the ND as used in this study is not a minority and does not refer to a particular ethnic group but to a group of people that inhabit the oil-bearing regions of Nigeria[32]. The argument is two-fold. The first is that the ND is in conflict with other ethnic groups in Nigeria for denying them access to their God-given natural resource. On the face of it this may seem plausible. But a closer reading of the ND situation will reveal the falsity of this argument. The people of the ND have never in their struggle singled out any ethnic group for vilification. Their anger has been directed at institutions like the military, the Nigerian government, the oil companies and their sub-contractors.

The second fold of the argument is that it is the question of a minority being denied their rights within a federation. As Osaghae put it, "their leaders attribute this injustice to the fact that they are minorities."[33] Put differently, the minorities are being oppressed by the dominant groups. To understand this let us bear in mind that the minority issue in Nigeria is a relative phenomenon. First, there are minorities in the north and even within the ND there are minorities. According to Osaghae "neglect and underdevelopment are not restricted to the Ogonis and oil-producing minorities, they are the objective condition of most groups in the country (though the point often made by oil-producing communities is that because they produce the bulk of the country's revenue they deserve special privilege)."[34]

If this is the case why is that of the ND different? It is not different; it simply exposes the limitations of the minority rights argument. In the north, the Fulani and Hausa are two different ethnic groups but in Nigerian political parlance, they are identified together as the dominant group.

This is not to suggest in anyway that the ND has been well treated by the other ethnic groups in Nigeria. This is far from the fact. What is at stake here is not how the rest of Nigeria has treated the ND but how, or the kind of relationship that exists between the people of ND and the oil companies and the modern state of Nigeria. What is it in their relationship that makes it inherently conflictual? That is the challenge. Most of the conflicts in the ND are internal – that is within the ND itself.

It has also been suggested that the main issue in the ND conflict is self-determination[35] for the people. This view misses the point because it ignores that fact that some of the conflicts in the ND are intra-community, interpersonal and even at times intra-personal[36]. If self-determination were the solution, perhaps the ND states of Delta, Rivers, Bayelsa and Akwa Ibom for instance would have been havens of peace. For instance, the Ogoni's allege domination by the Ijaws in Rivers State.[37]

The point is that self-determination, greater autonomy and independence may breed new minorities and new conflicts. Moreover, self-determination does not mark the end of class conflicts, relational conflicts and conflicts over status, values and resources.

The conflicts in the ND have also been positioned as human rights violations[38]. Human rights violations are the consequences of responses to conflicts. They are not the explanations of conflicts. This is more so in the ND. First, in the whole of Nigeria, Lagos area is believed to have experienced more incidents of human rights violations than any other part of the country[39]. One of the reasons for this has been due to excessive urbanization and the accompanying degree of very high social anonymity. In spite of these Lagos has experienced fewer intergroup conflicts compared to the ND – at least going by various reports[40].

As a matter of fact the positioning of the ND conflicts as human rights violations was a strategy by the indigenes and activists to draw attention to their plight. This is because since the collapse of the Soviet Union, the whole notion of self-determination or secession has ceased to excite the international community. This strategy has paid off because many human rights NGOs have been monitoring the situation in the ND. For instance Human Rights Watch, a foremost human rights group, since the hanging of the Ogoni Nine publishes an average of two reports per annum on the ND. According to Welch, "Ogoni demands illustrate the interplay of politics, economics and ethnicity within a context increasingly shaped both by access to international media and by human rights concepts".[41]

Like most other human activity, there are unintended consequences of seeing the ND issue as mere human rights violation. First is that after creating the awareness, what next? For instance, I have read at least ten different reports on the Umuechem incident mentioned in the last chapter. But surprisingly no human rights groups have tried to do something about the bounced check, which was issued to the survivors as compensation, especially given that the issuing of a bounce check is a criminal offence in Nigeria. None of the reports even mentioned the issue of the bounced checks.

Second, human rights conscientisation goes with responsibility otherwise those conscientised will become unmanageable. And this may lead them to equate lawlessness with human rights. According to Agboola and Alabi:

"As is to be expected when people feel that their families' health is at stake, they sometimes take actions that violate community standards of acceptable behaviour. Notably they engaged in civil disobedience and they occasionally even break the law"[42].

This is one of the explanations for major human rights violations in the ND. Many youths embark on some actions in the mistaken belief that they are fighting for their rights not knowing that they are committing a crime (see the introduction to the book). Even where crimes have not been committed these youths are sometimes framed by law enforcement agents in order to extract maximum punishment and to deter others from engaging in such actions in the future. Human rights violations therefore are neither causes nor explanations for the conflicts in the ND. Human rights violations are the effects or fallouts of the conflicts in the ND. If the conflicts are resolved and transformed, the issue of human rights violations will be minimized but that will not stop it entirely.

Most NGOs working in the ND have triple mandates. They combine environmental issues[43], with minority and human rights mandate. So apart from seeing the ND conflict as human and minority rights issues, they also interpret it as environmental degradation. First, these three do not go together because they all require different kinds of intervention. For instance, the government or company department that deals with human rights violations will not be the one to deal with environmental degradation. And different skills and tools are required to engage in each.

Wherever oil or other minerals are prospected the environment is affected. There are guidelines for organizations and individuals to follow. But whether they comply or not is not of much concern at this stage of the discourse. The question here is: if the environmental degradation is managed will it bring an end to the conflicts in the ND? The answer is of course no. The issue of compensation is still there. There is also resource control, self-determination, federalism, ethnicity and colonialism.

Again, environmental degradation is not the issue but a fallout of a mismanaged relationship. It is obvious that in our post-modern world, the environment has become a very important issue especially since after the Earth Summit in Brazil (1992) and the more current one in South Africa. Be that as it may, it is important to emphasize that this is by no means an attempt to minimize the impact of oil exploration activities on the ND environment. The ND has borne the brunt of the pollution that comes with oil exploration. But that is only a part of the problem. It is not the problem.

Wherever one form of production or the other is taking place, the environment is usually affected. An activity as "harmless" as farming destroys the environment. A good example is the Everglades in South Florida in the USA, where agriculture was fingered as one of the culprits in the damage. But the difference is that in the case of the Everglades[44] a 38 year and almost $7.8 billion Restoration Plan is in place. And this was put together with political will and participation of civil society.

Other authors emphasize bad constitution, class struggle, bad governance, failed state phenomenon, military dictatorship and lack of a democratic culture. Before examining each of this and the justifications for such explanations, it is important to note that conflict is a complex phenomenon and can never be explained from one perspective. However, since this study is not a mere academic excursion but a blueprint for intervention, it is toeing a pragmatic approach for intervention with visible entries and exits. Bearing in mind the above, George Klay Kieh Jr. wrote: "The central tenet of the eclectic theory[45] is that civil conflicts are the products of a confluence of factors – cultural, economic, historical, political and social. That is given the complexity of civil conflicts a single variable or factor is insufficient to explain the causes of these phenomena"[46]. So while one may not disagree with Kieh, it is also important to recognize the limitations of resources and skills during intervention.

Some other authors insist that the ND conflict is as a result of bad constitutions[47]. According to Professor Julius Ihonvbere, a "critical by-product of the new processes of dynamics of power, politics and political contestations has been a renewed interest in constitutionalism and constitution making all over Africa"[48]. People who subscribe to this view argue that since 1900 both the process and product of constitutions in Nigeria have been fraught with wrong-headedness. For instance they readily point to the process of making the constitution: "In essence the process that culminated in the 1999 constitution represented a deliberate perpetration of 'Political 419' on the Nigerian people and a betrayal of their yearnings for a transparent, accountable, just, and democratic political culture"[49].

First, reference is made to the opening statement in the preamble to the 1999 Nigerian constitution where it declares "we the people of Nigeria…"[50] They also make reference to section 55 of the 1999 Federal Constitution, which says that Hausa, Igbo and Yoruba (these are the three dominant ethnic groups in Nigeria) shall be the language of doing business in the national legislature. They also point to the 13% derivation formula for revenue allocation, which they derisively refer to as "87% deprivation".

The fact remains that no Nigerian constitution has been given a fair chance at implementation. Constitutions are mere guides to regulate human conduct. Constitutions are made for human beings and not human beings for constitutions. As Agboola and Moruf put it, "environmental law is a system of rules of social control which aim at creating order out of disorder in order to create a sustainable, liveable environment.

These laws mainly involve defining the relationship between human beings and the world they inhabit and the limits to which we may impact upon the environment..."[51] In chapter 2 of the same constitution, the *raison d'etre* of governance was clearly spelt out in the "Fundamental Objectives and Directive Principles of State policy". The same section makes good governance, democracy and social justice a mandatory responsibility of government. The issue then is not with the constitution but with the implementation.

The conflicts in the ND is not merely a constitutional issue since no constitution has been operated in Nigeria long enough to determine its suitability or otherwise. The former secretary general of the Commonwealth Chief Emeka Anyaoku put it thus: "Nigeria's problem at the time was not that of a faulty constitution, since no sincere attempt had been made to operate the 1999 constitution"[52]. If a constitution is bad, it could be amended. But when did Nigerians genuinely try using their constitution and found it wanting? People write constitutions and people implement them. For now all efforts should be geared towards implementing the new constitution with a view to smoothening the rough edges as, and when the need arises.

Too much effort shall not be devoted to the issue of class struggle because the conflicts in the ND affect all – irrespective of class. Emphasizing the class struggle argument is also a way of indirectly arguing that the ND issue is all about economics. Though oil as a resource is a money-spinner and the presence of oil may escalate a conflict, that does not necessarily mean that the conflicts there are only about economics. There is no doubt that there is class stratification in Nigeria. But that is not to say that particular groups of people are engineering the oil spills just to punish people of other class for instance. Or that a particular class is behind the non-payment of adequate compensation to land owners.

One is also aware that the Marxian class concept is more in-depth than the arguments posed above. However, to submerge the issue into class struggle will lead to gasping logic. I make this observation because both the elites and the common people are all involved in these conflicts. Class does not arise in the conceptualization of conflicts in the ND; it arises only in the sharing of the spoils from the conflict like compensation. As Peter Ozo Eson notes "several oil producing communities are known where privileged indigenes divert communal compensation from oil companies."[53]

The United Nations Development Programme (UNDP) defines governance "as the exercise of economic, political and administrative authority to manage a country's affairs at all levels, comprising the mechanisms, process and institutions through which that authority is directed"[54]. Good governance is a catch phrase that encapsulates most issues that have been raised above. Bad governance is the opposite. By bad governance we mean the inability of the state "to arbitrate between groups or provide credible guarantees of protection for groups."[55] Governance includes several other variables like the provision of food, shelter, education, and health for the people.

Taking cognisance of all these, one question comes to mind: if the Nigerian government guarantees all of the above for the people of the ND will the conflict there simply evaporate? One feels that an honest answer will be no. This is because there are more to the conflicts than the issue of basic human needs. There are issues of global human needs like respect, acknowledgement, recognition etc. If therefore the mere enthronement of good governance will not transform the conflicts, it means that bad governance is also just another aspect of the issue. More importantly governance is mainly concerned with erecting the structures or creating the enabling environment for a creative and non-violent approach to conflict resolution.

The concept of failed state[56] has also been brandished about as the cause and explanation of the conflicts in the ND. It seems however that most of those who make this argument use 'failed state' in an every day English language sense instead of the political concept that it is. A failed state is one where there is no centralized authority and the whole structures of governance have collapsed. Examples include Afghanistan before US invasion in 2001. Somalia is another good example. Sierra Leone was one before the peace process was brought back and kick started.

Nigeria has never been without a recognized authority. State structures might not be efficient but they exist and render service. May be the proponents of the failed state theory are referring to bad governance as well. For the avoidance of doubts, the ND have not lacked the presence of the apparatus of modern state, on the contrary, their complaint has been the "over presence of government security agents".

The failed state argument is therefore conceptually flawed. It does not also correspond to existing reality and therefore cannot be accepted as a credible explanation for the conflagration in the ND. After all the people of the ND are directing their demands at the institutions and representatives of the Nigerian state.

Military dictatorship and lack of democracy have also been blamed for the conflicts in the ND. The argument is that the conflicts in the ND have been exacerbated by the lack of democratic space to address the issues. Democracy however is not a magic wand that decrees conflicts out of existence. In fact in a seminal work just before he died in an air crash in 1997, Professor Claude Ake had argued that instead of the "collectivity, liberal democracy focuses on the individual whose claims are ultimately placed above that of the collectivity."[57] But the demand of the people of the ND is for increased participation in issues that affect them.

Though democracy may provide a possible route to explore intervention, the lack of democracy does not account for the conflicts in the ND.

It is also posited that the military has been responsible for the upsurge in the conflicts in the area. When the Willink Commission was set up in 1947 to address the issues of the ND, the military were not in power. When the Midwest region was created in 1963, the military were also not in power. When Isaac Adaka Boro and others declared the Republic of Niger Delta in 1969[58], it had absolutely nothing to do with the military being in power. In fact at that time Nigeria was still at war and Rivers and Cross River States had just been created.

To say that military rule accounts for the sufferings of the people of the ND is to miss the point. Searching the literature on the ND one could not point to any special policy thrust by the civilian regime of Shagari that was specifically targeted at the ND except for the Presidential Task Force which recommended that 1.5% of the federation account be devoted to the development of the ND[59]. Obasanjo's "thank you" visit to the people of the ND for voting massively for his party was to upbraid them after ordering the destruction of Odi. Two reports are instructive for us to see how the people of the ND feel about civilian regimes. In fact the title of Human Rights Watch Report "The Niger Delta: No Democratic Dividend" says it all.

Democracy is dictatorship of the majority and a crude game of numbers. Lack of democracy and practice of democracy all impact the dynamics of conflict in one way or the other. It is true that inherent in a democratic set up are in-built conflict resolution mechanisms. However, these mechanisms do not operate themselves. Human beings manage them. One feels that anchoring the ND issue around democracy was also another strategy to create awareness considering the wave of democratization that was sweeping across the world. This use of the democratic chance to politicize discontent as Ted Robert Gurr[60] put it has its own unintended consequences.

First, it pitched the ND movement as an opposition political movement instead of an affirmative action initiative. Since the military were trained to see enemies and allies, the ND people were seen as enemies to be done away with. Second, it portrayed the agitating groups in the ND as political movements rather than social movements. Finally, it created a partisan colour for a movement that would have retained its credibility by remaining politically non-partisan. I use the term non-partisan advisedly.

Another very important view that is pervasive is the "oil causes conflict argument". A recent headline in *Thisday* newspaper screamed "Crude Oil, Source of Instability – Governor Rashidi Ladoja." In the report, the governor "described the presence of crude oil in any location in the country as a source of instability to the region. The prospecting of crude oil had become a source of conflict and ready-made tool of tribal and religious upheaval. He regretted that rather than be a source of multiple blessing for Nigeria, the presence of crude oil has now been associated with ethnicity and instability in the nation"[61]. Governor Victor Attah of Akwa Ibom (an oil producing state in the ND) also expressed similar views in a newspaper report[62].

In a recent study titled "Does Oil Hinder Democracy?" Professor Michael Ross of the University of California, Los Angeles, concluded that "the oil-impedes-democracy claim is both valid and statistically robust; in other words, oil does hurt democracy"[63]. In a similar report on the conflict in Sudan titled "God, Oil and Country: The Changing Logic of War in Sudan" the International Crisis Group (ICG) wrote, "A prime reason why the north has always resisted southern separatism is the latter's natural resource wealth. Particularly central to the current equation are oil"…[64] The World Bank (WB) supported this view thus: "Statistically, secessionist rebellions are considerably more likely if the country has valuable natural resources, with oil being particularly potent".[65] To support this conclusion, the WB report argues that it was the discovery of oil in Nigeria that motivated the Igbos to rebel against the Federal Government of Nigeria.

There is no doubt that oil may have acted as an added incentive for Biafra to secede, but why did the Federal Government of Nigeria not allow Biafra to go after they captured the oil fields in the early part of the war? The oil (resources) explanation may have been given prominence for two reasons. First, a bank that sees every thing in terms of money and resources commissioned the report. Second, the report was "to alert the international community to the adverse consequences of civil war for development".[66] Since the report was aimed at the consequences of conflict, its content will not be very useful for explaining the dynamics of conflicts.

However in his report, Ross adds this caveat "there is nothing inevitable about resource curse".[67] Ross' thesis is not that oil causes conflict, rather that the presence of oil may breed a carefree attitude of government towards the citizenry because government does not depend on tax revenue, and therefore will not be accountable to the people. No matter where governments get money from, accountability is a sine qua non for good governance. It has nothing to do with the presence of, or lack of oil or any other resources for that matter. This helps buttress the argument that it is not laws or structures or resources that is the issue in the ND but the issue is around people and relationships.

In another report commissioned by Oxfam and written by Ross, the argument of resource curse was clarified. The report said that mineral dependent states tend to be corrupt, authoritarian and ineffective thereby committing an unusually high chunk of the mineral resources to defence and wars. The reason according to Ross is that "extractive sectors tend to be capital intensive and use little unskilled labour. Therefore, they do little to help the poor or the employment rate. Such industries are geographically concentrated and consequently create small pockets of wealth that typically fail to spread"[68].

Apart from oil, the World Bank (WB) added another issue to explain conflicts generally. Though this WB report is not on the ND but on civil wars generally, it is relevant to this study. "This report argues that civil war is now an important issue for development. War retards development, but conversely development retards war".[69] Its relevance lies in the fact that other scholars share the same view that the problem in the ND is due to lack of development. Development is a very complex phenomenon. For the purposes of this study we shall take development to mean the provision of the basic needs of life. It shall also mean the provision of security etc. It is interesting that the same WB report concludes on page 5, "but neither will development secure global peace". The report concludes that "economic development is central to reducing the global incidence of conflict; however, this does not mean that the standard elements of development strategy – market access, policy reform, and aid – are sufficient, or even appropriate, to address the problem"[70].

Conflicts in the ND are therefore relatively independent of the variables enunciated above. However, a change in the above mix, like any other social dynamic, will definitely affect the conflicts. What has happened is that most authors wrote from a given perception or lens. More often than not this was either from a professional or vocational viewpoint or from the point of view of a sympathizer. Having dismissed some of the explanations of the conflicts in the ND, the next task is to examine what is really the conflict in the ND. This will be done after laying the theoretical foundation of the conflicts in the ND. The next chapter will address this.

Notes and References

---

[1] See for instance "Boiling Point" (2000) by Committee for the Defense of Human Rights.

[2] See Boiling Point, pp. 197-205.

[3] A prominent member of this school is the venerable professor of economics, Eskor Toyo (ibid: 6-8)

[4] Some of them include Walter Rodney's How Europe Underdeveloped Africa, Chinweizu's, The West and the Rest Of Us and Madubuike's, Towards the Decolonization of African Literature.

[5] See Osarhieme B. Osadolor, "The National Question in Historical Perspective", in Momoh, A and Said Adejumobi. The National Question in Nigeria: Comparative Perspectives. Burlington: Ashgate Publishing Company, 2002, p.38.

[6] See Festus Iyayi, "Oil Companies and the Politics of Community Relations in Nigeria", in Boiling Point: The Crises in the Oil Producing Communities in Nigeria. Lagos: Committee for the Defence of Human Rights, 2000, p. 176.

[7] See Ikime, Obaro. The Fall of Nigeria: The British Conquest. London: Heinemann Publishers, 1982.

[8] See Crawford Young, "The Heritage of Colonialism", in Harbeson, J.W. and Donald Rothchild (Ed.). Africa in World Politics: The African State System in Flux. Boulder, Colorado: Westview Press, 2000, p.37.

[9] See Dibie, R. Understanding Public Policy in Nigeria: A 21st Century Approach. Lagos: Mbeyi & Associates, 2000, p.4.

[10] See Whitaker, C.S. "A Coda on Afrocentricity" in Richard L.Sklar and C.S. Whitaker (eds.) African Politics and Problems in Development, Boulder, Colorado: Lynne Rienner Publishers, p.357, 1991.

[11] See Achebe, C. The Trouble With Nigeria. London: Heinemann, 1983, p.1.

[12] See Awa, E.O. Issues in Federalism. Benin: Ethiope Publishing, 1976, p.41.

[13] See Mwakikagile, G. Ethnic Politics in Kenya and Nigeria. Huntington, NY: Nova Science Publishers, 2001, p. 4.

[14] See Environmental Rights Action (ERA). The Emperor Has No Clothes. Port Harcourt: ERA, 2000, p.7.

[15] Ibid: p.7.

[16] Ibid; p.7.

[17] Mark Anikpo, "Social Structure and the National Question in Nigeria", in Momoh, A and Said Adejumobi (Ed.). The National Question in Nigeria: Comparative Perspectives. Burlington, VT: Ashgate Publishing, 2002, p.61.

[18] See 2000 Annual Report on the Human Rights situation in Nigeria by Committee for the Defence of Human Rights, p.256.

[19] Leith, Rian and Hussein Solomon, "On Ethnicity and Ethnic Conflict Management in Nigeria", African Journal of Conflict Resolution, No.1, 2001, p.3.

[20] Donald Rothchild and J.W. Harbeson, "The African State and State System in Flux", in Harbeson, J.W. and Donald Rothchild

(Ed.). Africa in World Politics. Boulder, Colorado: Westview Press, 2000, p.7.

[21] Akinyemi, B. (Ed.). Readings on Federalism. New York: Third Press, 1980, p.29.

[22] For a detailed analysis of the state creation exercises see Osaghae, E.E. Crippled Giant: Nigeria Since Independence. Bloomington: Indiana university Press, 1998.

[23] See Godfrey N. Uzoigwe "Assessing the History of Ethnic/Religious Relations", in Uwazie, E.E. et al (Ed.). Inter-Ethnic and Religious Conflict Resolution in Nigeria. Lanham, MD: Lexington Books, 1999, p.16

[24] See Awa, E.O. op. cit.

[25] Uwazie, E.E. et. Al (Ed.). Inter-Ethnic and Religious Conflict Resolution in Nigeria. Lanham, Maryland: Lexington Books, 1999.pp.36-48.

[26] Sam G.Amoo. "The Challenge of Ethnicity and Conflicts in Africa: The Need for a new Paradigm. UNDP, 2003, p. 3.

[27] Thomson, A. An Introduction to African Politics. London: Routledge, p.58.

[28] Edward E.Azar, "Protracted International conflicts: Ten Propositions", in Burton, J and Frank Dukes. Conflict: Readings in Management and Resolution. New York: St. Martin's Press, 19990, p.146.

[29] Kruger, P. (Ed.). Ethnicity and Nationalism: Case Studies in Their Intrinsic tension and Political Dynamics. Marburg: Hitzeroth, 1993, p.11

[30] Bostock, W. Language Grief: A "Raw material" of Ethnic Conflict, in Nationalism and Ethnic Politics. 1994, Vol. 3, No. 4, pp. 94-112.

[31] Osaghae, E.E. "The Ogoni Uprising: Oil Politics, Minority Agitation and the Future of the Nigerian State". African Affairs (London), 94, 376, July 1995, pp. 328-332

[32] See "Boiling Point." P.10.

[33] Ibid; 328.

[34] Ibid; p.331.

[35] See Tebekaemi, Tony (Ed.). Major Isaac Jasper Adaka Boro: The Twelve-day Revolution. Benin City: Idodo Umeh Publishers, 1982, p.68.

[36] See Douglas, O. and Doifie Ola, "Nigeria: Defending Nature, Protecting Human Dignity – Conflicts in the Niger Delta", in Searching for Peace in Africa, 1999, www.euconflict.org.dev/ECCP.

[37] Osaghae, E. E. Ogoni Uprising. P. 332.

[38] See Human Rights Watch Report, "The Niger Delta: No democratic Dividend", Vol. 14, No.7, (A) October, 2002. See also Constitutional Rights Project Report titled, "Land, Oil and Human Rights in Nigeria's Delta Region", 1999.

[39] See CDHR Annual Reports 2000, 2001, 2002.

[40] Babawale, Tunde. The Rise of Ethnic Militias, De-legitimization of the State and the Threat to Nigerian Federalism. West Africa Review, Vol.2, No.1, 2002.

[41] Welch, C.E., Jr. "The Ogoni and Self-determination: Increasing Violence in Nigeria". The Journal of Modern African Studies, 33, 4 (1995), pp.635-649.

[42]Tunde Agboola and Moruf Alabi, "Political Economy of Petroleum Resources Development, Environmental Injustice and Selective Victimization: A Case Study of the Niger Delta Region of Nigeria", in Agyeman, Julian et al (Ed.) Just Sustainabilities: Development in an Unequal World. Cambridge, Massachusetts: The MIT Press, 2003, P.283

[43] Ibid: pp. 267-288.

[44] See Pryor, B. Ethnographic Study in Environmental conflict Resolution: The Role of Environmental NGOs in the Florida Everglades Restoration. (An unpublished PhD dissertation), 2003, submitted to the Dept. of Conflict Analysis and Resolution, Nova Southeastern University, Florida

[45] For a detailed description of the Eclectic Model Of Conflict, see Fisher, R.J. The Social Psychology of Intergroup and International Conflict Resolution. New York: Springer-Verlaag Publishers, 1990.

[46] Kieh, G.K. Jr. and Ida Rousseau Mukenge (Ed.). Zones of Conflict In Africa: Theories and Cases. Westport, CT: Praeger Publishers, 2002, p. 12.

[47] For a detailed view of the people of the ND on the 1999 constitution See the Report titled "The Emperor Has No Clothes: Report of Proceedings of the Conference on the Peoples of the Niger Delta and the 1999 Constitution. Port Harcourt: Environmental Rights Action, 2000.

[48] See CDHR. State Reconstruction in West Africa. Lagos: CDHR, 2001, p.16.

[49] ibid: p.25.

[50] Constitution of the Federal Republic of Nigeria, 1999, p.1.

[51] Tunde Agboola and Moruf Alabi, op. cit. p.282.

[52] See Thisday on Sunday, January 18, 2004.

[53] See CDHR. Boiling Point, p.50.

[54] UNDP Report, "Promoting Conflict Prevention and Conflict Resolution Through Effective Governance", 2003, p.5.

[55] Lake, D.A. and Donald Rothchild, "Containing Fear: The Origin and Management of Ethnic Conflict", International Security, Vol.21, No.2, 1996, pp. 41-75.

[56] For peacebuilding paradigms in a failed state see Heinrich, Wolfgang. Building the Peace: experiences of Collaborative Peacebuilding in Somalia (1993-1996). Life and Peace Inst. 1997.

[57] Ake, Claude. The Feasibility of Democracy in Africa. Dakar, Senegal: CODESRIA, 2000, p.10.

[58] Tebekaemi, Tony (ed.). Major Isaac Jasper Adaka Boro: The Twelve Day Revolution. Benin City: Idodo Umeh Publishers, 1982, p.68.

[59] See NDDC profile 2002, p.4.

[60]See Gurr, T.R. Why Men Rebel. Princeton, NJ: Princeton University Press, 1970.

[61] See www.thisdayonline.com/news/20031224news05.html

[62] Daily Champion, Saturday, September 14, 2002, p.3.

[63] Ross, M.L. Does Oil Hinder Democracy?.World Politics, 53, April 2001, p.356.

[64] See ICG. God, Oil and Country: The Changing Logic of War in Sudan. Brussels: ICG Press, 2002, p.99.

[65] World Bank: Breaking the Conflict Trap: Civil War and Development Policy. Washington: World Bank and Oxford University Press, 2003, p.60.

[66] ibid: ix.

[67] ibid: p.357.

[68] See Thisday Newspaper, Wednesday, October 17, 2001, p.38.

[69] ibid: p.1

[70] Ibid: p.6

Chapter Four

Conflicts in the Niger Delta: A Theoretical Framework

In the previous chapter we examined some of the explanations proffered for understanding the conflicts in the Niger Delta. Many were dismissed either for being conceptually flawed, limited in scope, inadequate for analysis or doctrinaire. This chapter will look at the nature and dynamics of conflicts in the ND with a view to establishing some theoretical foundations for analysing them. The following shall guide the discussion in this chapter. Conflicts in the ND fit exactly into Peter Coleman's notion that conflicts have "an extensive past, a turbulent present and a murky future"[1].

It is important to clarify the use of conflicts in plural instead of the singular. This is because there are many conflicts in the ND[2]. There has been this mistaken assumption that there is only one conflict in the ND. As this chapter will show there are a multiplicity of conflicts, parties, issues and contexts. Any theoretical analysis that ignores this fact will not be very useful for a clear understanding of the conflicts in the ND.

Another very important issue is whether the conflicts in the ND are social or political conflicts. Because conflicts are dynamic social phenomena, it is not always useful to box them into one category. But in this case it is important because "modes of resolution are fundamentally related to the nature of the conflict"[3]. Rather than use the categories of social, civil or political conflict, it is better to argue that the conflicts in the ND are not the equivalent of wars. "To be classified as a civil war, a conflict must pass a certain threshold, producing a certain number of combat-related deaths (usually one thousand) over some period of time (usually one year)"[4]. With these simple criteria, the ND conflicts do not qualify as war. This can be justified with several reasons. First, there are no standing armies fighting the cause of the ND. Second, there are no battlefronts. For instance, the ND people recognize individual's role as an indigene and as a staff of government or oil company. Earlier on, we dismissed the ethnic conflict argument. For the purposes of this study, the conflicts in the ND are characterized as social conflicts, because, they are embedded in the social-interaction process. But they have far reaching implications.

First, conflicts arose in the ND because the people are in relationship with other people and institutions. Mack and Snyder agree with this when they observed that, "conflict has been characterized as a basic social-interaction process"[5]. For instance, the people of the ND are in relationship with the oil companies, governments etc. Second, the people of the ND were not "prepared" for what was coming. In the mindset of the elites of the modern state of Nigeria, since they control the infrastructure of violence, every one shall be coerced into compliance. This was based also on the mistaken assumption of the self-righteous and unifying mission of the Nigerian state. Agboola and Alabi captured it this way:

"...thus the development of oil resources in Nigeria is a joint responsibility of the state, multinational corporations, multilateral organizations and local elites. Oil resource development, according to these actors, means economic growth. Implicit in their actions is the justification of the prevalent neo-liberal belief that economic growth is absolutely good and that its benefits ultimately trickle down to everyone"[6].

Closely related to this is the non-recognition of the people of the ND as the "natural owners, nurturers and custodians"[7] of the resources in the ND. The conflict in the ND is a manifestation of a deep yearning among the people to be respected, recognized and acknowledged not only for their ownership of the resources but for bearing the brunt of its exploitation. Third, the disagreements that breed the conflicts in the ND are issues that challenge the core identity and values of the people of the ND. This shall be elaborated on later. Four, the non-renewable nature of the exploitation of the resources of the ND leaves a scary prospect of extinction in the minds of the people of the ND. The people of the ND believe albeit correctly that the government and oil companies are "liquidating its own natural capital, rather than living off its harvest".[8]

Five, there is a conflicting worldview[9] as to what is happening in the ND. This came out clearly from the literature review in the previous chapter. This shall be elaborated when we discuss the responses to the conflicts in subsequent chapters.

Six, the history of resource management and control[10] in the Nigerian nation, to the people of the ND, smacks of selective victimization. For instance, it was not until after Arab-Israeli (Yom Kippur) war of 1973-74 that oil became the major foreign exchange earner for the Nigerian nation. Before this time each region controlled its own resources. But with the rise in the price of oil, the Nigerian state took over absolute control of the oil industry. Oil, unlike cocoa, groundnut and palm oil is unique because it is capital intensive and requires sophisticated technology. People in the other regions of Nigeria could relate to the production of food and cash crops because it has been part of their life. Government bought their products through the various commodity marketing boards[11]. The oil industry is not organized that way. More on this later when we examine in depth causes of the conflict in the ND.

Seven, there is a fluidity in the identity of parties and issues in the conflict in the ND. Before the advent of colonialism the people of ND wanted to be able to play the role of middleman in the slave trade. At another time they wanted to trade directly with the companies in Liverpool. At other times the issues fluctuated from "restructuring Nigeria's federalism, to self-determination to resource control". This also applies to the parties. There seem to be no consensus among the parties as to what is going on and what is to be done. For instance, an oil company executive once lamented, "The problem in the ND is that of proliferation of ammunitions".[12]

In another study Professor Augustine Ikelegbe[13] identified close to one hundred different associations that are parties to the conflict in the ND. The implication of this is that there are about one hundred viewpoints, entry points and exit points. And more importantly this implies one hundred different constituencies to engage.

Finally, there is what one may call "hierarchy of conflicts" in the ND. These conflicts shall be identified in the later part of this chapter. The important thing is that each conflict must be resolved and transformed before the next. This is because each one has a linkage to the one before it. The diagram below illustrates how each of the conflicts is related to one another. For instance, without understanding and resolving the various land disputes, it will almost be impossible to address the chieftaincy disputes. This is because it is only when the issue of who owns the land is settled that that of who governs it will make sense. The same thing applies to the issue of leadership of Community Development Associations (CDA). Disagreements between elders and youths revolve around the issue of representation. This is directly related to the issue of oil spills. Oil spills have a direct impact on land, compensation and leadership.

The point is that most interventions have either lumped the conflicts as one or identified the issues and parties as one. Categorizing the conflicts as LeVine did, as "structural levels of conflict"[14] will be misleading. First, the conflicts below do not refer to levels but to types. One could have land dispute between families, communities and even individuals. The essence of arranging the conflicts in a hierarchical order is to gain a bird's eye view of the conflicts in the ND. This is important so that when intervening we will know exactly which of the conflicts we are working on. Again, conflicts in the ND do not necessarily have such clear-cut demarcations. An oil spill might occur in a land that is in dispute. In that instance, the skills and training and the resources available locally will play a great role in intervention.

Moreover, this hierarchical arrangement of conflicts fits perfectly into Lederach's pyramid of actors because it helps us to see who is doing what and at what level. It also ties with Dugan's description of conflicts into four different categories, and of Docherty's three worlds model of conflict. And it gives us an idea of the timeframe for each intervention because with Lederach's time dimension of peacebuilding we are able to plan our interventions properly.

Hierarchy of Conflicts in the Niger Delta

The conflicts in the Niger Delta can be ranked in descending order of intensity as follows:

1. Self-determination
2. Resource control
3. Citing of developments projects
4. Oil spill
5. Conflict between elders and youths

6. Leadership tussle in associations
7. Chieftaincy tussle
8. Land dispute

Types of conflicts identified in the Niger Delta

Before delving into the details of the theoretical framework, a few comments on the structure of the oil industry will illuminate the discussion. In Nigeria, the federal government controls the oil industry through the Nigeria National Petroleum Corporation (NNPC). The NNPC came into being in 1977. Its progenitor was the Nigeria National Oil Company, which was founded in 1971 in compliance with the Organization of Petroleum Exporting Countries (OPEC) Resolution Number XVI. 90 of 1968. OPEC was founded in 1960 and was formally constituted in 1961. OPEC members produce about 40% of the world total crude oil. The eleven members possess more than three-quarters of the world's total proven crude oil reserves. Nigeria joined OPEC in 1971 while Venezuela suspended her membership of OPEC in 1992 and Gabon terminated her membership in 1995. Its headquarters is in Vienna, Austria.

The resolution referred to earlier directed member countries to "acquire 51% of foreign equity interests and to participate actively in all aspects of oil production"[15]. With this resolution, the people of the ND had thought that they would have a say in how the oil in their land is exploited. This was not to be. The essence of this discussion is to show that the Nigerian government and oil companies are not the sole determinants of what happens in the oil industry. This will affect any intervention proposed and even the nature of the conflicts.

The NNPC has many subsidiaries. They include the Pipelines and Product Marketing Company (PPMC), Directorate of Petroleum Resources (DPR), and many others. The NNPC is involved in both upstream and downstream operations. These include exploration, production, refining, pipelines and storage terminals, marketing of oil, gas and refined products and petrochemicals. The government appoints major personnel in the NNPC structure. As government own agency, the NNPC is also involved in conflicts with host communities[16].

The NNPC runs what is known as Joint Venture Operations (JV) with other multinational oil corporations in Nigeria. These multinationals include Shell, AGIP, Chevron, ELF, Mobil and Texaco. Over the years the NNPC has continued to reduce its stake in the operations of the oil companies. For instance, recently it lowered its stake in Shell from 80% to 55%. The reason for this may not be unconnected with its inability to meet up with its financial obligations under the joint venture agreements.

The license to participate in any kind of activity in the oil industry is granted by the Federal Government of Nigeria. The Oil Minerals Act of 1969 invests the federal government with absolute control of all minerals, including oil within the territory of Nigeria. This includes both onshore and offshore. This control is reinforced with the Land Use Act of 1978, which invested the ownership of all lands in Nigeria in the governments. "Prior to this decree, land was communally owned and the various traditional rulers, clan heads, and community leaders had the power to determine customary law insofar as this affected land tenure and use"[17]. The land use Act[18] made the people of ND tenants in their own homeland to an institution that they neither recognize nor are able to understand or relate to.

The implications of the above are many. First, the oil companies have no legal[19] obligations whatsoever towards their local host communities except in the case of oil spill. Second, the host communities have no say absolutely in how and what happens to oil revenue. Third, most of the legislations regulating the oil industry were enacted under military regimes, the import of these were not lost on the people of the ND. The 1969 legislation was to break the backbone of Biafra during the Nigerian Civil War 1967-70, while the Land Use Act was meant to whittle down the influence of traditional rulers. "No where else in Nigeria has the impact of the Land Use Decree manifested, in all its imperfections and inequities, as in the Niger Delta region, Nigeria's main oil producing region"[20].

However, the most important implication of the existence of the NNPC is that it is a big bureaucracy, corrupt and inefficient. Unfortunately, the oil companies cannot undertake any development project without approval from the Directorate of Petroleum Resources (DPR) of the NNPC. According to an oil company staff "it takes the DPR an average of twelve months to approve one borehole project and before they do, the budget has gone up by almost 100%"[21]. The point is that the oil industry in Nigeria, which is synonymous with the NNPC, is highly politicized. It is not run for profit but for political patronage.

Oil is no longer a mere local resource. The price, production and distribution of oil are tied to the international system. According to Michael Fleshman of the New York-based Africa Fund, "the oil producing communities in the Niger Delta, while often remote, isolated and inaccessible, are inextricably linked to the international community through the production and sale of oil"[22]. So the communities watch as things happen to them and they have neither say nor control over issues around the oil industry. We shall see how this has impacted the dynamics of conflict in the area.

Under these joint venture agreements the oil companies pay royalties, rent and tax to the federal government. After this they do not feel obligated to commit any further finances towards development in their host communities. But because of the near absence of government presence in these communities, the oil companies are now playing a surrogate role or alternative government. "Over time the companies have often become the effective local governments in the areas in which they operate"[23]. They complain that this is increasing the cost of doing business, while the communities claim that it is their right to enjoy the benefits of God's gift to them. This is at the core of the conflict in the Niger Delta. A recurring question among the people of the ND is who owns the land? Second, who bears the bulk of the impact of oil exploration activities? Is the government responsive and responsible enough to use the resources from the area to develop the region? How should funds from oil revenue be allocated? Which should be given greater weight in the allocation - population or derivation?

The relevance of this brief overview of the structure of the oil industry in Nigeria is that it reinforces my earlier assertion that the government is not in a position to act as a credible third party intervener in the conflicts in the ND. The NNPC is supposed to regulate the oil industry, but it is also involved in oil production activities. So it cannot regulate since it is engaged in the same activity. Because of its tie to government, it cannot also mediate because it is not credible. Moreover, it has not shown the required competence and professionalism to act as a credible third party in the conflicts in the ND.

My approach will not be the traditional one of analysing the causes of conflicts under such sub-heads as economic, social, political and psychological. The approach will be using specific examples to point out the exact instances where the misunderstandings that led to the disagreements began. This approach will not involve an in depth analysis of the case studies. Rather it will draw out relevant data to illustrate specific issues. Second, this analysis will concentrate on specific conflict issues and not looking at the conflicts in the ND as a whole.

The major challenge with this methodology is: can the lessons learnt be generalized for other conflicts, and in different communities? The aim is not to propose a generally applicable theory, but to provide a framework for analysis and intervention in conflicts in the ND. The argument is that the main reason for the failure of most interventions is because of defective analytical methodology. For instance, many of the workshops on the ND on conflict resolution incorporate mediation. The question is mediation between who and whom? Second, what are the criteria for selecting participants for these workshops? Are the participants being trained to become conflict resolution practitioners or to understand their own conflicts or intervene in conflicts in their communities?

The first issue for analysis is land dispute. The guiding theoretical framework here will be the Robert Ardrey's "territorial imperative"[24]. Ardrey has argued that we as human beings, just like animals, deserve and defend a territory where our basic needs and interests are met. These needs include security, food, identity and prestige. He concluded his treatise by arguing that we defend such territories at all costs from those who try to undermine these interests and needs. This clearly captures the land conflict in the ND. Conflict erupts when the claims of the people of the ND to land and territory become incompatible with the desire of the government or the oil companies or other communities or individuals to satisfy their own basic interests and needs within the same physical territory.

To understand the causes and sources of land disputes in the ND, one needs to take a look at the land tenure system[25]. The land tenure system in the ND is both complex and complicated. In the ND, land is collectively and individually owned. Land is owned collectively by the community or family and individually by a person. Land is used for three main purposes: for farming, residential, burial and other sacred purposes. Land is owned mainly by adult males who at death bequeath same to their male children.

If someone wants to build a house for instance, he summons the community together, gives them a goat, cooks food and he is given land to build. The community rejoices that another son of theirs have become a "man". The community even assists him in building the house. At death, our man's son inherits the compound. If he does not have a son, the land reverts to the community. But should the deceased son decide to get his own place, he loses his right over his father's land and the land reverts to the community. More importantly land was a renewable natural resource. Land was also sustainable through shifting cultivation and multiple cropping.

Then came the modern state of Nigeria. Land was no longer managed by the community. It was managed and controlled by an unknown outside force (the government). The Land Use Act of 1978 effectively put the ownership and management of land and the resources in it into the hands of government. As Williams put it, "the land use Decree was designed to pose a direct challenge to alternative sources of societal authority by relegating all private transactions in land to governmental agencies"[26] "Prior to this decree, land was communally owned and the various traditional rulers, clan heads and community leaders had the power to determine customary law in so far as this affected land tenure and land use"[27].

First, everyone wanted to keep what he or she has. Second, private ownership of land was to be documented (e.g. certificate of occupancy or power of attorney). Third, the most important evidence of ownership is occupation. Four, disputes arising from land was now to be settled by those who did not even understand the traditions of the community (the courts).

Five, the more one paid, one could acquire as much land as possible. All these were happening against the backdrop of a growing population. For instance "as the growth of oil business turned Warri into a land of opportunities, people flooded Warri from all parts of the country in search of the 'golden fleece'. The result was to heighten the stake for Warri land and the bitter contest by the three major ethnic groups"[28]. There are therefore three dimensions to the land dispute in the ND as illustrated by the above example. First, was that land became scarce due to the influx of foreigners. This raised the cost of land and rent. Second, it increased the value of land and subsequently the intensity of disputes over land between individuals and families. Third, it also meant increased tension between ethnic groups within the same locality.

This meant that land was no longer a renewable resource. Second, government was to decide whether land would be used for agriculture, for oil production or for any other purposes. There was a fresh influx of non-indigenes to the area, which meant a joint ownership of land. The import of this was that land and its sacredness became demystified. Since, the people did not have a say in the oil industry, they have lost control over their land. And they could not farm, and since the only land that was respected is one that is occupied, people struggled to acquire and occupy as much as they can since it was the only thing permanent.

The question then is this: why is land dispute as intense as it is in the ND? The reason is simple. It is about survival. The struggle over land in the ND is the struggle not only for a means of livelihood, but a struggle for life. Second is that all the traditional institutions that mediated land disputes have given way to modern ones. These modern ones do not hold the same awe and aura among the people. For instance part of the mechanisms for resolving land disputes was oath-taking[29]. It was believed among the people that if one swore falsely to an oath claiming a piece of land that did not belong to him that the person would die. This is not the same situation with the modern methods of resolution, which is formal, logical and argumentative. Moreover, the traditional method is timely and instant while the modern method could last for a whole lifetime[30].

Related to the question of land dispute are the issues of compensation for land, and the purchase of land. Among the people of the ND, land was not sold outright. But with the arrival of oil companies and various legislations, they began to pay compensation for crops and economic trees on the land and not for the land itself. The main issue is, to whom or for what would compensation be provided? This is in view of the complexity of the land tenure system discussed above.

The first question is how much compensation is enough for a piece of land that will be bequeathed to an oncoming generation? Second, if you pay compensation now, what when the money is exhausted? These were issues that mere legislations could not address. The misunderstanding was that when the oil companies paid compensation, they thought that they had acquired the land in perpetuity. They made the locals sign all kinds of documents that meant and signified nothing to them. To the people of the ND land is legacy that is held in sacred trust for an oncoming generation and not an expendable resource.

Another aspect to the land issue is when communities and families and sometimes individuals begin to contest for a piece of land. This usually gets intense if oil has been found on the land or when the piece of land in question provides access to an oil installation. This distinction between hosting oil installation and providing access to oil installation is very important[31]. This is because the oil companies and in fact government insist on doing business with what they refer to as host communities. But some communities that provide access to those installations even suffer more damage. This complicates the whole issue of land the more.

Since the mechanisms for resolving the conflicts are unreliable, the only option which the disputants foresee is to eliminate the other. And when this is done they now have a free reign over the disputed land. An example is the land dispute between the Eleme and Okrika people over who owns the land where the Port Refinery is built.

So in the ND land per se does not cause conflicts. What causes the conflict over land in the ND is that the communities have not come to grips with the social change that have affected their land tenure system. And no one has been able to 'prepare' them for it. There is no doubt that the presence of oil may intensify these conflicts. But the main cause of the conflict over land in the ND is the perception of the kind of change that has taken place over land issues in the area. This is how Oronto Douglas captured this:

> "The forests of the Niger Delta were exploited for timber, wildlife and plant life, our cultures and social dynamics were `studied' by all sorts of people: 'scholars', journalists, 'travellers'. It has been suggested that not all these studies were for the sake of knowledge and science. We began to

imbibe things that were not in our constitutional character. 'Forests' became the derogatory 'bush'. Religion became 'fetish', 'paganistic', 'idolatry' and our people's culture and way of life became 'primitive', 'uncivilized'. Our nations were reduced to 'tribes' and our industries like gin brewing became 'illicit'. The governance of our land was subjected to political authorities that were far away, unaccountable, not transparent and very brutal and arrogant"[32].

This is very important and needs further elaboration. In the other regions of Nigeria like Western, Eastern and Northern, land was used for agricultural purposes namely for the production of cocoa, palm oil and groundnut respectively. The people produced these commodities on their farmlands, while the government bought the produce through the various marketing boards[33]. This meant that the people still had a say over how the land was managed. They also participated in the economic activity that took place on the land. This was also not a rentier culture.

This is not the case with the ND. The oil industry is a technical one that is capital intensive. The people of the ND are not playing any role because oil exploration has rendered them mere rent and compensation collectors. And unfortunately this rent and compensation is not even always paid. And when they are paid, they are not usually sufficient. All these intensified the struggle for land, especially given the exponential growth in population.

The next important conflict issue in the ND is that of chieftaincy tussle. This aspect of the conflict in the Niger Delta is hardly mentioned in the literature. The reason for playing down chieftaincy disputes is that many ND indigenes and scholars[34] generally feel that the disputes are fuelled by the oil companies to divide the communities[35]. John Burton refers to this as the scapegoat theory[36]. This is a situation where internal conflicts are blamed on some external foes. This however cannot fully explain the recurrent conflicts in the ND. Moreover oil companies have been quick to deny these allegations[37]. Chieftaincy disputes in Nigeria are real and intense. To understand this, let us put it in context.

Before the arrival of the Europeans many communities were ruled by kings and chiefs. In communities where there were no kings elders governed the place with the assistance of the chief priests. After the amalgamation in 1914, the British still made use of these chiefs. In the south it was known as indirect rule[38]. Before the amalgamation, the chieftaincy institution was a very strong and revered one. The process of selection was thorough and rigorous; this is because the kings combined political, judicial and spiritual functions. There were checks and balances. The kings ruled with the help of a council of chiefs or elders in council. The chiefs and kings drew their legitimacy directly from the people. As Professor Nwabueze puts it, "a chief obtains his title and authority by virtue of rules of succession which derive their binding force from their acceptance by the people"[39].

The importance of the chieftaincy institution is also underscored by the fact that they were among the brains behind the slave trade, and later what was called legitimate trade. But more importantly, "the chiefs were the authorities from whom British jurisdiction in Nigeria was acquired… and most of the treaties by which this jurisdiction was ceded recognized the right of the ceding chiefs to continue to rule their people"[40].

Professor Ikime has also documented the prominent role of chiefs in the Niger Delta in the resistance to foreign domination of their economy. Many paid dearly for their role in this resistance. However one thing that is important which Ikime pointed out was the rise of slaves to the various ruling houses. After the abolition of the slave trade, the ex-slaves became prominent in the trade in palm oil. "The question which arose was whether the delta society could prevent these wealthy and therefore powerful slaves from seeking to wield political authority with their economic strength"[41].

All these were to change the status of the slaves. There was no other place where the change was more profound than in the Niger Delta. There are two important issues to be noted in this development. First, it meant that the traditional institution was open for contest. Second, it meant that it could also go to the highest bidder. This changed the criteria for selection of traditional rulers. It also de-legitimised the claims of the various ruling houses. In other words the institution became politicised. This raised the stakes for the office of traditional rulers. "Of important significance in the disturbances in the oil producing areas is the role of traditional rulers"[42]. This prominence was to be reinforced with the a provision in the Land Use Act "that compensation for surface rights over land acquired for oil activities would be made to traditional rulers to disburse as they deemed fit in their tradition on behalf of their communities"[43].

From the foregoing, a few points need to be emphasized. First, is that the chieftaincy institution is a very important and influential one among the indigenes of the Niger Delta. Second, this prominence has been reinforced by various governments namely the British, the military and the politicians. Third, the traditional institution comes with a lot of privileges and with the advent of oil; succession disputes will become more intense. Four, with the modern state of Nigeria tinkering with the traditional institution, occupiers of such thrones will definitely be cultivated by government. In fact governments began to influence who should occupy the thrones.

Chieftaincy disputes in the Niger Delta played a prominent role in the Warri Crisis[44]. It also featured in

Ogoni, Okrika[45] and Kula Kingdom[46]. When the Federal Government of Nigeria established the Presidential Panel on National Security (PPNS), one of the areas of concern was chieftaincy disputes[47]. In fact, in the ND there is an association known as Traditional Rulers of Oil Mineral Producing Communities in Nigeria (TROMPCON)[48]. It is instructive that this organization emphasized oil minerals instead of minerals generally since most communities that have other extractive resources experience the same kind of environmental problems like the oil-bearing communities. There are two possible explanations for this. First is that oil is the mainstay of Nigeria's economy as at now. Second is that this is an attempt by these traditional rulers to gain visibility. They also spoke for their various communities.

From the foregoing it is obvious that chieftaincy disputes are not peculiar to the Niger Delta but are prevalent all over Nigeria. The question then is why is it such a big issue in the ND? Why are chieftaincy disputes so violent and intense in the ND? Why must it constitute an important area of intervention in resolving and eventually transforming the conflicts in the Niger Delta? What has been done to transform these chieftaincy tussles? How do chieftaincy disputes fit into the overall picture of conflicts in the ND?

One could seek for possible explanation in the "resource curse" argument[49]. This resource curse thesis posits that countries with abundance resources end up being poorer and therefore unstable than those without resources. This is not to suggest that these resources are a curse but that the presence of these resources are often a catalyst for the escalation of various kinds of conflicts, since it makes the government less responsive, and less accountable to the people. If we translate that to the ND situation, it means that since the traditional rulers could easily make money from oil companies and various government agencies, they are prepared to do anything to ascend that throne since they do not need legitimization from the people.

Another form of conflict in the ND that is closely related to the above is that of leadership of Community Development Associations (CDA) or other civil society groups. Civil society groups are part and parcel of ND society. They have farmers, hunters, and fishermen's guilds. They also have age sets and other types of associations of various titleholders. Professor Augustine Ikelegbe[50] has documented these associations. A key point to note is that there is a proliferation of organizations in the ND.

Many reasons account for this. First, it is historical. The people have organized themselves around associations since time immemorial. According to an Umuechem Community Development Plan, a community development document put together by Shell, Umuechem, a community of about 4,000 with twelve families has eight different community-based organizations. This is apart from the Community Development Committee and the Central Union. Second, is that the oil companies prefer to deal with groups rather than individuals. Third, this is a collectivist society where group harmony and solidarity hold sway. Fourth, the individual's worth is measured through his membership of the community. Five is that oil companies have a strategy of working through development associations. Six, is that the there have been an upsurge in the number of NGOs since the end of the Cold War. The explanations for this shall not detain us. Moreover as Welch, Jr. observed the demands of the people of the ND have been shaped by human rights and environmental concepts[51].

The question therefore is why is it that the leadership of these organizations that are meant to serve the people have become a do or die affair? Again in the ND positioning is a critical factor if one is to benefit from the oil largesse. Leadership of such organizations means access to those in power both in government and in the oil companies. It also provides access to funding especially for NGOs. Many individuals have risen from community to national leadership positions. Though there is nothing wrong with this phenomenon, the issue remains that it has led to unbridled competition for positions.

Ordinarily occupiers of these positions are supposed to serve the people but in the ND just like elsewhere in Nigeria, this has not always been the case. Because of the privileges that accrue to these leaders, the struggle becomes so intense and violent. Another possible reason is that since being an ordinary citizen handicaps one from benefiting from the oil largesse, positions become an avenue to partake in the sharing of the loot.

In one community for instance, the leader of a CDA had wanted the contract for the installation of concrete electric poles for himself; because of this he did not inform his community that the oil company in question had advertised the contract for bidding. When we wanted to inform the community about this progress, the so-called leader came to warn us that it was not advisable for us to visit the community because according to him there was trouble there. We defied him and went to the community.

On getting there, the people were waiting for us to update them on the progress, which we had made in negotiating their demands with the oil company. When this community leader learnt that we were in the community, he fled with some members of his executive. The move to replace him was so intense that we had to intervene to avoid a break down of law and order[52].

Another very critical issue is that these community associations are not usually formal organizations. Many are without constitutions; many do not know what the terms of officeholders are. There are no institutional mechanisms for checks and balances. And anyone who could put together a group of people forms an organization and begins to make demands on government, oil companies and even their communities. In fact I was informed in one community that most of these organizations have metamorphosed into some kind of ethnic militias. The radicalisation of these fringe organizations is also an issue that has affected the dynamics of conflicts in the ND[53].

Examples abound of many organizations in the ND that have been ripped apart by leadership tussle. The Movement for the Survival of Ogoni People has just managed to come out of its own internal leadership squabble. In fact the leadership crisis started before Ken Saro Wiwa and others were hanged in 1995. Even after this, the trouble continued when Ledum Mitee took over the leadership of the organization. It was so intense that Mitee decided to step aside[54]. In many other communities, there was evidence of similar conflicts.

I think one of the most important explanations for the leadership crises in most ND organizations is the issue of ideology. Let me elaborate on this a little. In the ND, there are two broad trends as to how the people of the ND should respond to the issues affecting them especially as it relates to their relationship with the oil companies and government. For the purposes of this study we shall categorize them as liberals and radicals.

The liberals believe that they should consistently and constantly engage the oil companies in dialogue and debate while the radicals believe that enough is enough. Many traditional rulers belong to the liberal camp. Others in this camp are politicians, top government officials and oil company executives and businessmen. The radical group is made of up human rights activists, journalists, youths, students, workers, women and the peasantry.

Both groups have very strong justifications for their various positions. So in every community, the leadership and in fact the entire community, is divided along these lines. In Ogoni for instance, one group referred to the other as vultures[55]. This is a metaphor for those they claim are feeding on the flesh of Ogoni people. The radicals are the new comers and are trying to take over the leadership of the ND. They believe that the liberals have been too compromising. At times they even accuse the liberals of being corrupt. So the radicals have a messianic complex of having been ordained to save the ND. So when the leadership of any organization in the ND is at stake, these two tendencies compete for power. In many instances the outcome has been very bloody. The story of Ogoni is too familiar to warrant recounting.

Another dimension to the conflict is the struggle between the traditional rulers and the leadership of the CDAs. The issue often has been that of who speaks for the community. Many youths and so-called progressives believe that the traditional rulers consort with the government and the oil companies. This has led to what they term a betrayal of the ND cause. They see the traditional rulers as too conservative and compromising.

They represent "the emergence of a new generation of educated youths who were now only openly challenging corrupt chiefs and community leaders; but were also successfully mobilizing their communities…"[56] This tussle between the traditional rulers and leaders of CDAs is one that has adversely affected the entire conflict dynamics in the ND.

Inter-communal conflict is another kind of conflict that has bedevilled the ND over time. These conflicts could be over land, compensation, leadership, the citing of community development projects and many others. Within Ogoni these conflicts have claimed more than 1000 lives[57]. Several Ogoni villages were sacked between July and September by members of the Andoni community, by the Okrika in December 1993 and by Ndoki in April 1994[58]. In another community, the Federal Government intervened to restore peace and order[59].

Why do communities in the ND fight themselves? There is no special reason since this is a general phenomenon in Nigeria - in fact all over the world. The main issue has been land. Many communities struggle over land. In one instance, a community has claimed that an oil company paid compensation to a community that does not produce oil. In this clash twenty people were reported to have died in Olumaibiri in Bayelsa State[60]. In other instances clashes between communities have been as a result of government policies like the location of a local government headquarters or the citing of a development project. Some communities like Owaza do not want others to benefit from the oil largesse since they claim that they host oil installations[61].

Another conflict that is very critical in the ND is the conflict over oil spill. In 2003 Shell Nigeria reported 221 oil spills. The report claimed that 80 of the spills were as a result of equipment failure while the remaining 141 were as a result of sabotage. But the World Bank says that "oil spills are generally caused by companies themselves, with corrosion being the most frequent cause."[62] Shell's 2003 Annual Report also said that 9,900 barrels of crude oil was spilled. But activists from the ND have consistently disputed these figures. In an article recently Torulagha dismissed the above figures from Shell in these words:

> "...oil spillage continues to be a major problem in the region. Contrary to the rosy picture painted by the Anglo-Dutch Company, (SPDC), the indigenes and the environment continue to suffer from oil spillage"[63].

Oil pipelines pass through the length and breadth of the ND. In some communities these pipelines are buried in the ground, while in others they are on the surface[64]. The law is that when oil spills from these pipelines, that the oil company will pay compensation according to the damage. However, there is a proviso. If the oil spill is as a result of sabotage, the oil company will not pay but if the spill is as a result of equipment failure, then the oil company will pay. The complicated nature of this whole thing about oil spill makes it inherently conflictual. I will try to put this in context.

The Oil Pipelines Act of 1956 makes provision for the payment of compensation arising from oil exploration activities, while the Petroleum Act of 1969 makes a list of such items on which compensation would be paid. They include economic trees, structures on land, shrines and venerable objects. Further the 1999 constitution makes provision for the payment of adequate compensation. The Petroleum Production and Distribution Anti-Sabotage decree of 1975 deals with the issue of sabotage and wilful damage of oil production facilities. This decree provides for the death penalty or 21 years imprisonment. To reinforce this law, the military government enacted the Miscellaneous Offences Decree No. 20 of 1984, which also prescribes the death penalty for sabotage of oil installations[65].

The first conflict over oil spill arises as to whether it is sabotage or equipment failure. This is usually established through a joint investigation team comprising the oil company, affected communities, and the Department of Petroleum Resources (DPR) of the Nigerian National Petroleum Corporation (NNPC). The first problem is that the oil company in question hires and pays the expert that investigates the cause of the spill. He who pays the piper dictates the tune – as the saying goes. Often the DPR pleads lack of funds or expertise to join in the investigation. The World Bank in fact noted that the DPR was "not able to perform its duties and is limited to obtaining oil company spill reports."[66] The communities are mere watchers. If the experts return a verdict of sabotage, the community will disagree. And if they return a verdict of equipment failure the oil company will dispute it[67]. That is the first part of the conflict.

The second part of the conflict is, if it is sabotage, who did it? The communities have argued on several occasions that even if oil spill was as a result of sabotage, that it does not necessarily follow that someone from the community did it. The oil companies have countered by saying that the communities should be able to know who did it since it happened in their community. The communities counter that they are not security agents or law enforcement agencies, and that even if they knew the culprits of such sabotage, they doubted the oil companies would accept their verdict on the matter.

Another aspect of this whole affair is that oil spill could be on land or in the river. If it is on land, it spreads until the spill is contained. If it is on the river, the water carries it from the point of spill across so many communities. The issue then is if the spill is as a result of sabotage, do you still deny the victims the compensation due to them as a result of the spill?

It is important to elaborate more on this sabotage issue. It is believed that communities in the ND deliberately damage oil pipelines to cause spills in order to claim compensation from the oil companies. But the question which interveners in the ND have refused to ask themselves is this: Considering the massive environmental impact of oil spills, why would individuals deliberately pollute their own environment?

To the oil companies, this is caused by greed - the need by the communities to make money from the oil companies. The argument is that farming and fishing constitute subsistence form of economic activity and the people of the ND get some bulk payment from oil spill compensations. Because of this they sabotage equipments. This argument, though plausible on face value, lacks legitimacy when subjected to critical scrutiny.

In the first place, greed and grievances are not the only explanations for conflict behaviour. It is important to situate people's actions in a conflict context in their proper perspective. Kriesberg has critically examined the processes of conflict escalation. Sabotage of oil installations within the ND context is not the conflict itself but a response to conflict. "As the cause becomes more valued, ever more harmful acts are justified."[68] Osaghae says that "the significance of the letter to the companies was that it showed the people's loss of confidence in the state".[69] In another page Osaghae continues that "the main redressive mechanisms for oil-producing communities in their demand for improved conditions and better treatment have been petitions and delegations to the federal and state governments as well as the oil companies"[70]. Osaghae therefore concluded that it is the failure of the government and oil companies to respond to the demands of the oil-bearing communities that leads them to react the way the do.

Even in the area of litigation, where the communities still stand at a disadvantage, Frynas argues that "the suppression of litigation against oil companies has important consequences for conflicts in the oil producing areas. Difficulties in obtaining legal recourse may have contributed directly and indirectly to informal forms of conflict such as seizure of oil industry equipment or the kidnapping of oil company staff".

Continuing Frynas wrote that "the resulting paradox is that the companies' ability to stifle one form of protest may lead to different, perhaps more troublesome forms of protest."[71] He concludes by arguing that "oil companies have contributed significantly towards discontent in the oil producing areas. Oil companies could contribute towards resolving conflicts in several ways: by reducing the adverse impact of oil operations, by ending over-reliance on security co-operation and by executing meaningful development projects in tandem with the local people. Unless the government and the oil companies change their basic attitude towards the local people in the Niger Delta, conflict and litigation are there to stay."[72] I shall discuss this basic attitude which Frynas made reference to in the above quote.

As we have seen, sabotage of oil company installations, which results in oil spills, is one of the responses of the communities to the conflict in the ND. They sabotage the equipments to signal their frustrations of not being listened to by both the oil companies and government. It is also an indication of a lack of confidence in the conflict resolution mechanisms instituted by the government and the oil companies.

Another aspect of the oil spill issue is that of cleaning the spill. Communities accuse oil companies of carrying out poor cleaning activities. In one site, which I visited in Ogoni, I was told that the oil company set fire to the spill. At Ogbodo, I was shown the spill site, which the company claimed to have cleaned[73]. Oil companies also plead security concerns as part of the reason why they do not clean up on time, and why the cleaning is sometimes not properly done. Oil companies also allege that the communities complain of poor clean-ups when the clean-up contract is not awarded to them, and also do a bad job when such contracts are awarded to them.

The communities counter that the main cause of poor clean-up is corruption. It is difficult to establish how much of this is due to the complexity and nature of the oil industry. What Frynas did for instance in his analysis of oil spills was to ferret out the little information he could get from court cases and judgments. I think that there is still a lot to be done in investigating the effects of oil spill on the dynamics of the conflicts in the ND. This will take time because it is at the heart of one of the enduring legacies of corruption.

Suffice it to mention that conflicts over oil spills and oil pipeline explosion have caused a lot of violence and bad blood in the ND. In 2003 security agencies in Nigeria impounded 19 vessels used for oil bunkering. Legislations have proved inadequate in dousing the flames of this very conflict. This does not mean that no intervention will work. It only means that before an intervention is put in place, we really need to understand fully what the issues are.

There is also the issue of boundary disputes between states in the ND. These disputes again are not peculiar to the ND. Boundary disputes are common all over Nigeria especially with the frequent creation of states. What often makes the dispute more intractable is the presence of oil on the disputed land. For instance the states get some extra money from the number of oil wells located in their areas of jurisdiction. Since 1976, Imo and Rivers states and later Abia have been embroiled in a dispute over some 43 oil wells in their boundary.[74] Usually these conflicts affect communities that live in and around these boundaries and they begin to fight against each other. This is another aspect of the conflict in the ND that needs to be looked at separately and not lumped together under the omnibus conflicts in the ND.

The other conflicts in the ND include that of resource control and on/offshore oil dichotomy. Resource control has to do with the demands of people of the ND to control their resources (oil and gas) as it is done elsewhere. They point to Sheffield in England and many other communities where the local communities that bear the brunt of the production of oil and enjoy enhanced revenue status.

The on/offshore dichotomy is a very recent addition to the conflicts in the ND. This became pronounced when the Federal Government of Nigeria under Olusegun Obasanjo decided that oil explored in Nigeria's coastal waters will no longer be credited to states but directly to the Federal Government. Since these two are constitutional and legal matters, it shall not detain us. Moreover, I strongly feel that even if these legislations are passed or the courts determine the whole issue of on/offshore oil dichotomy as they have done, it will also not remove the other conflicts discussed above. However, it was reported that president Olusegun Obasanjo has signed a bill into law abrogating the onshore/offshore oil dichotomy.[75]

Notes and References

---

[1] Peter Coleman. "Intractable Conflict", in Morton Deutsch and Peter Coleman (ed.). Handbook of Conflict Resolution. San Francisco: Jossey-Bass, 2000, p.432.

[2] This was one of my findings when I researched this project.

[3] Mack, R.W. and Richard C. Snyder, "The Analysis of Social Conflict – Toward an Overview and Synthesis", Conflict Resolution, Vol.1, No.2, June 1957, p.238.

[4] See Ross, Michael. How Does Natural Resource Wealth Influence Civil War? Los Angeles, CA: UCLA, 2001, p.8.

[5] Op. cit. P.238.

[6] Agboola, Tunde and Moruf Alabi, "Political Economy of Petroleum Resources Development, Environmental Injustice and Selective Victimization: Case Study of the Niger Delta Region of Nigeria", in Agyeman, J. et al (Ed.) Just Sustainabilities: Development in an Unequal World. Cambridge, Massachusetts: The MIT Press, 2003, p.274.

[7] See Cable, S. and Cable, C. Environmental Problems: Grassroots Solutions: The Politics of Grassroots Environmental Conflict. New York: St. Martins Press, 1995, p. 107.

8 Op. cit., p.278.

9 For a detailed analysis of how worldviews impact conflict see Docherty, J.S. Learning Lessons From Waco: When the parties bring their gods to the negotiation table. Syracuse, New York, 2001.

10 See "Meandering Pains of Resource Control" by Prof. Mark Nwagwu in The Channel magazine, vol.2, no.10, November 2001, pp.16-19. See also "Resource Control an Answered Prayer" by Funke Aboyade in Thisday Newspapers, www.thisdayonline.com/law/20021001law02.html

11 See H. Laurens Van Der Laan, "Marketing Boards and Colonial Trading Companies", The Journal of Modern Africa Studies, 25, 1 (1987), pp.1-24. See also William O. Jones "Food Crop Marketing Boards in Tropical Africa" the Journal of Modern Africa Studies, 25, 3 (1987), pp.375-402.

12 This was in one of the Stakeholders' Workshop held in 2002 in Port Harcourt. The name of the executive shall remain anonymous.

13 See Augustine Ikelegbe, "Civil Society, Oil and Conflict in the Niger Delta Region of Nigeria: Ramifications of Civil Society for a Regional Resource Control", in Journal of Modern African Studies, Vol. 39, No. 3, 2001, pp. 437-469.

14 Robert A. LeVine, "Anthropology and the Study of Conflict: An Introduction", The Journal of Conflict Resolution, Vol.5, No.1, March 1961, p.4.

15 See Okonta, I. And Oronto Douglas. Where Vultures Feast. San Francisco, CA: Sierra Club Books, 2001, p.55.

16 Interview with Wisdom Dike a Niger Delta indigene and founder of Community Rights Initiative an NGO that was involved in the conflict between the Egi community and the NNPC.

17 Okonta, Ike and Oronto Douglas. Where Vultures Feast. Op cit. p.26.

[18] For a detailed analysis of the Land Use Act, see Udo, R.K. Land Use Policy and Land Ownership in Nigeria. Lagos: Ebieakwa Ventures Ltd, 1990.

[19] For a detailed study of the legal aspect of the conflicts between the oil companies and communities see Frynas, J.G. Oil in Nigeria: Conflict and Litigation between Oil Companies and Village Communities. Hamburg: LIT, 2000.

[20] See Constitutional Rights Project (CRP). Land, Oil and Human Rights in Nigeria's Niger Delta Region. Lagos: CRP, 1999, p.3.

[21] Interview with oil company staff in Port Harcourt who prefers anonymity. The interview was conducted in June 2002 as part of STEPS TO CONFLICT PREVENTION project of Collaborative for Development Action, (CDA), Boston, USA.

[22] Committee for the Defence of Human Rights (CDHR). Boiling Point: The Crises in the Oil Producing Communities in Nigeria. Lagos: CDHR, 2000, pp.179-195.

[23] Ibid: 179-195.

[24] Ardrey, R. The Territorial Imperative; A Personal Inquiry into the Animal Origins of Property and Nations. London; Collins, 1967.

[25] The data for this analysis was collected in 2002 in an interview with a Chief from Umuechem who prefers anonymity.

[26] Donald C. Williams, Measuring the Impact of Land Reform in Nigeria", The Journal of Modern African Studies, 30, 4 (1992), p. 587.

[27] Okonta, I. and Oronto Douglas. Where Vultures Feast. San Francisco, CA: Sierra Club Books, 2001, p.42.

[28] Imobighe, T.A. et al (Ed.). Conflict and Instability in the Niger Delta: The Warri case. Ibadan: Spectrum Books, 2002, p.xii.

[29] Interview with a community leader from Ogbodo community who prefers anonymity.

[30] Frynas, J.G. Oil in Nigeria. P.171.

[31] A chief from Ogoni who prefers anonymity told me this in an interview when I was collecting data for this project in July 2002.

[32] Environmental Rights Action/Friends of the Earth (ERA), Nigeria. The Emperor Has No Clothes. ERA/FOEN: Benin City, 2000, p.141.

[33] See H. Laurens Van Der Laan, "Marketing Boards and Colonial Trading Companies", The Journal of Modern Africa Studies, 25, 1 (1987), pp.1-24. See also William O. Jones "Food Crop Markeing Boards in Tropical Africa" the Journal of Modern Africa Studies, 25, 3 (1987), pp.375-402.

[34] See Okonta, Ike and Oronto Douglas. Where Vultures Feast. San Francisco, CA: Jossey-Bass, 2001.

[35] Ledum Mitee the president of MOSOP alluded to this before the Oputa panel

[36] Burton, J.W. Systems, States, Diplomacy and Rules. London: Cambridge, 1966, p.124.

[37] Thisday, Friday February 2nd, 2001.

[38] Isichei, Elizabeth. A History of the Igbo People. London: Macmillan Press, 1976, p.141.

[39] See Nwabueze, B.O. A Constitutional History of Nigeria. London: Longman Publishers, 1982, p.93.

[40] Ibid: 92

[41] Ikime, O. The Fall of Nigeria: The British Conquest. London: Heinemann Publishers, 1982, p.18

[42] Constitutional Rights Project. Land, Oil and Human Rights in Nigeria's Delta Region. Lagos: CRP, 1999, p.26.

[43] ibid: 27.

[44] Ibid: pp. 26-27.

[45] Sunday Champion, November 25, 2001, p.7

[46] I interviewed a young man who was detained for his role in the Kula Kingdom saga in Port Harcourt in July 2003.

[47] Thisday, November 28, 2001, p. 13.

[48] Daily Champion, Saturday, September 14, 2002, p.3.

[49] Gelb, Alan. Oil Windfalls: Blessing or Curse. New York: Oxford University Press, 1988. See also Auty, R. M. Sustaining Development in the Mineral Economies: The Resource Curse. London: Routledge, 1993.

[50] Journal of Modern African Studies, Vol.39, No.3 (2001), pp. 437-469.

[51] Claude E. Welch, Jr. The Ogoni and Self-determination: Increasing Violence in Nigeria. The Journal of Modern African Studies, 33, 4 (1995), pp. 635-649.

[52] Details of this could be gotten from the Center for Social and Corporate Responsibility in Port Harcourt. The community and those involved shall remain anonymous for security reasons.

[53] Osaghae, E. E. "Ogoni Uprisng". P.333

[54] Thisday, Friday January 5, 2001.

[55] This was used several times to refer to the Ogoni Four who were murdered when I was intervening in Ogoni.

[56] Okonta, Ike and Oronto Douglas. Where Vulture Feast. Sierra Club Books: San Francisco, 2001, p.144.

[57] See the documentary Delta Force produced by Independent Television, Channel 4, Ireland.

[58] Claude E. Welch, Jr. The Ogoni and Self-determination: Increasing Violence in Nigeria. The Journal of Modern African Studies, 33, 4 (1995), pp .635-649.

[59] Thisday on Sunday, January 21, 2001.

[60] Thisday on Saturday, February 24, 2001.

[61] Interview with an indigene of Asa in July 2003 while I was researching this project.

[62] Frynas, J. G. "Oil in Nigeria." P.165.

[63] See The Niger Delta, Oil and Western Strategic Interests: The Need for an Understanding by Priye S. Torulagha (www.unitedijawstates.com/niger_delta.htm

[64] In July 2002 I visited Ken Saro Wiwa's father in the company of a delegation from the Ecumenical Council on Corporate Responsibility from London, we were shown these surface pipelines in and around the compounds.

[65] For a detailed analysis of Petroleum legislations in Nigeria see Frynas, Jedrzej Georg. Oil in Nigeria: Conflict and Litigation between Oil Companies and Village Communities. Hamburg: LIT, 2000.

[66] World Bank 1995, Vol.II, annex J, quoted by Frynas J.G. "Oil in Nigeria." p. 86.

[67] The Center for Social and Corporate Responsibility based in Port Harcourt, Nigeria conducted a detailed investigation into the Batan Oil spill. Their findings are on a video. I was part of the team that conducted the investigation.

[68] Kriesberg, L. Constructive Conflicts. P.161

[69] Osaghae, E. E. "Ogoni Uprising", p. 337.

[70] Ibid: p.332

[71] Ibid: p.226.

[72] Ibid: p.231.

[73] Details of these interventions are available at the CSCR office in Port Harcourt.

[74] Thisday on Saturday, February 24, 2001.

[75] Thisday August 6, 2004.

Chapter 5

Responses to Conflicts in the Niger Delta

In the previous chapter we discussed the eight different kinds of conflict that were identified in the ND during our field investigation. These conflicts include land, chieftaincy, community leadership tussle, oil spill and pipeline explosions, compensation, citing of community development projects, on/offshore oil dichotomy and resource control. In this chapter we shall discuss the various responses to the conflicts identified in the previous chapter. This shall be done from four main perspectives namely: government, communities, oil companies and NGOs. The main purpose of this is to enumerate how these various parties have responded to these conflicts. The importance of this is for us to know what worked and why, and what did not work and why. It will also help us to know what has been done, so that we do not duplicate an already existing initiative. Finally, this chapter will give us insights into how each of the parties understands the conflicts or rather how each of the parties interpret the conflicts.

A very important observation from these responses is that for the eight different kinds of conflicts identified there have been different responses from each of the parties. This is a problem. For instance, it would have been interesting to see the oil companies, government, communities and NGOs work collectively and collaboratively on such issue as land. But this has not been the case. The issue is that if all the parties are trying to solve a specific problem why can't they work together?

Response to conflicts could be direct or indirect. By direct I mean a situation whereby the parties confront themselves to address the conflict. Indirect is where one of the parties approaches a third party to assist with an intervention. Responses could be violent or nonviolent. This is self-explanatory and would not delay us.

Responses could also be verbal or non-verbal. For instance, many people respond to conflicts by simply avoiding them while some will simply keep mum or refuse to talk about it. Responses could also be formal or informal. By this I mean that sometimes groups of people or individuals may decide to formally present their grievances in formal settings. Others may prefer informal channels for resolving their conflicts. Frynas[1] refers to such acts as hostage-taking and vandalisation as informal, while such methods as negotiation he refers to as extra-legal. Responses to conflicts could also be collective or individualistic. Though a conflict may involve a whole group of people, one or two may decide to take their destiny into their own hands. Finally responses to conflicts could be short term or long term.

In discussing the responses to conflicts in the ND we can see that all the above mentioned models of responses were used in many of the cases. The importance of discussing these different models is that each response comes with its own unintended consequences. Second, it will be interesting to find out why people choose a particular type of response instead of another. Third, will be why certain kinds of conflicts recommend certain types of responses. But one thing that will be clear as we discuss responses to conflicts in the ND is that the over-riding consideration in choosing a particular model of response is result expected, and frustration. When people get frustrated they change to another model. Finally, it would appear that most of the people did not sit down to make a conscious choice of their responses. The responses appeared to be chosen at random or what I may refer to as trial and error basis. This is where civil society groups or third party interveners would have proved very helpful to the ND by educating them on the various types of responses and the likely outcomes.

The other issue is that some responses are suited for certain contexts. For instance knowing the notoriety of military regimes, it would not have been advisable for the people of Umuechem to pour onto their streets. May be that strategy would have worked better under a civilian regime. On the other hand, it would have been very fruitful if people were also educated on when and how to respond to conflicts. For instance, one feature of responses to conflicts in the ND is that almost all of them are reactive and not proactive. There has been very little emphasis on what Lederach[2] and Curle [3] call 'proventive peacebuilding'. These shall be discussed more in the next chapter.

Another important thing to note in discussing responses to conflict in the ND is that most of the parties have used litigation extensively for all the eight different conflicts identified in my investigation. The use of litigation was based on the assumption of the effectiveness of legislations. There are so many laws regulating the oil industry in Nigeria. But the use of litigation has been hampered by lack of funds, lack of confidence in the judicial process and general frustration[4]. Finally we need not bother how effective each of these responses have been for the evidence speaks for itself. But we shall concern ourselves with why they were not effective.

Government response

As far as the government is concerned the issue in the ND is that of security. In discussing government responses to conflict in the ND it is important to distinguish between military dictatorships and civilian regimes. This is important because both regimes approach issues from different mindsets.

Under the immediate post independence civilian regime of Sir Abubakar Tafawa Balewa, the conflicts in the ND were not as intense and violent as they became in the 1990s. The main issue then was the agitation for the creation of states. Eventually the Midwest region was created in 1963. The creation of the Midwest region had more to do with the need to whittle down the influence of the Action Group (the dominant political party in the western region) than resolving conflicts in the ND. The same 'privilege' of new states was not extended to the eastern part of the ND.

It could be argued that the conflicts in the ND became more pronounced after the civil war (1967-70). Many reasons have been adduced for this. But two very significant incidents stand out. First, was the declaration of the Republic of the Niger Delta by Isaac Adaka Boro in February 1966, and second was the creation of states in 1967. During and after the war the ND suffered unduly, but it also dawned on the people of the ND that the oil, which the other major ethnic groups were fighting over, was their 'property'. Further, "the solidarity of the inhabitants initially drew upon a sense of grievance engendered in the Nigerian civil war during 1967-70, when many non-Igbo members of the

The Republic of Biafra felt discriminated against. The Ogoni was particularly hard-hit. Their oil-producing areas were occupied by Nigerian forces in 1968, and a great many of the inhabitants were forcibly moved into the Igbo heartland. Indeed, Saro Wiwa later estimated that 30,000 Ogoni had perished, "over ten percent of the total population of the ethnic group"[5]. There are some theoretical explanations why conflicts generally escalate. This will be discussed later. Meanwhile we shall concentrate on the responses.

It was in the early 1990s that the ND literarily caught fire. Probably the most significant trigger was the establishment of the Movement for the Survival of Ogoni People (MOSOP) and the involvement of the cerebral Ken Saro Wiwa. A series of events quickly followed. First the Ogoni cultural organization *Kagote* gave birth to MOSOP. MOSOP anchored her agitation around the concept of rights, self-determination, environmental integrity and ethnic minority issues.

To give vent to the above MOSOP drafted the "Ogoni Bill of Rights". In my opinion, it was the first time that the ND issue was thoroughly articulated. With this, MOSOP found an international audience. Saro Wiwa's reputation as a writer was to prove extremely useful to the Ogoni cause in particular and ND in general. From that time events moved at a dizzying speed. There was the Ogoni National Anthem, Ogoni National Flag and finally was the show down with the Nigerian government in the 1993 elections[6].

MOSOP networked with such organizations as Unrepresented Nations and People Organization (UNPO) based in Holland, The New York-based International Federation for the Rights of Ethnic, Linguistic, Religious and Other Minorities, Amnesty International, Human Rights Watch and so many others joined the fray. The internationalization of the conflicts in the ND was also synonymous with the politicization of the conflicts in the ND. This changed the whole dynamics of the issues at stake.

The most enduring implication of this was that all other ethnic groups in the ND took a cue from Ogoni and began to mobilize and organize. Two things happened simultaneously. While the ND organized government responded with a heavy hand[7]. In May 1993 the military government promulgated the Treason and Treasonable Offences Decree, which prohibited, with possible death penalty, the agitation for ethnic autonomy.

Let me try to situate the government's response a little bit in detail, and within the context of the evolving political scenario of Nigeria. First the June 12, 1993 election had just been held. It will be recalled that the election was later annulled. The ND as represented by MOSOP had boycotted that election. Those who called for the Ogoni to participate were labelled vultures[8]. Second, the Ogoni were calling for secession. Third, they anchored their agitation around issues of democracy and good governance. This was anathema to Nigeria's ruling military oligarchy. To them the only interpretation of the ND cause was opposition. This meant that they should be dealt with.

So the most prominent response of the government was a military solution to the demands of the ND. Instead of the regular police, an Internal Security Task Force was set up in January 1994. Unfortunately, as I write under a democratic regime, there is also the Joint Task Force for Peace in the ND[9]. Apart from this, the government has also responded in several other ways. The overriding consideration in government's response to conflicts in the ND is to suppress the conflicts with such a heavy hand that no other communities in the ND will ever agitate again. To what extent this has succeeded is obvious even to the amateur observer. If this is the situation, why has government not bothered to rethink their strategies for intervention? That will be discussed in the next chapter.

The colonial government set up the Willink Commission to look into the fears of the minorities. This commission did not recommend the creation of states instead it recommended constitutional guarantees to protect the rights of minorities. One major achievement of the Willink Commission was the establishment of the Niger Delta Development Board. This board was inherited by the civilian administration of Sir Abubakar Tafawa Balewa. The coup of 1966/7 killed this board.

Again were the creation of Midwest region in 1963 and the creation of states in 1969. It is interesting that as the government was ostensibly attending to the issues raised by the people of the ND, it was also consolidating its hold on the oil industry. For instance in the same 1969 when it created states the government also promulgated the Oil Minerals Act which invested in the Federal Government absolute control of all minerals including oil within Nigeria. The government followed this up with the Land Use Act of 1978, which again put all land in Nigeria under the firm grip of the Federal Government.

It was not all bad news. In 1992 the military regime of Babangida established the Oil Minerals Producing Areas Development Commission (OMPADEC) with Decree 23 of 1992. OMPADEC was to address the developmental needs of the ND. To ensure its success, an ND indigene Chief A.K. Horsfall was appointed its executive chairman. Perhaps the only memory of OMPADEC that is left with the people of the ND is the monumental corruption, which it has been reputed for.

This was to be succeeded by the Petroleum Special Trust Fund (PTF). The PTF was set up as a political survival strategy and not to address the conflicts in the ND. In 1994, after incessant fuel scarcity, the Abacha government decided to increase the price of fuel in order to combat smuggling. The increase was predicated on the premise that government was subsidizing the price of fuel for smugglers. In order to prove that he meant business, Abacha removed the so-called subsidy on the price of fuel. He decided to put the subsidy money in a special account to develop national infrastructure. The agency that was to manage this fund was named PTF. One of the major failings of the PTF was that it became an omnibus kind of government agency. Coupled with this was its Father Christmas kind of disposition. Needless to say that by the time that Obasanjo came in the PTF had outlived its usefulness.

Currently there is the Niger Delta Development Commission (NDDC). The NDDC is unique in conception because it makes provision for the joint funding of development projects in the ND between the government and oil companies. Basically the NDDC is a development agency. It has been carrying out several development projects in the ND. But the NDDC has not been resolving conflicts. In fact the NDDC's philosophy is still development dumping. Though the ND needs development, the question is, can we have development in a conflict-ridden environment? A more fundamental question is, is development a panacea for conflicts? These are some the challenges we shall grapple with when we try to redefine conflict resolution initiatives in the ND.

Apart from the above, there has been the tinkering with the revenue allocation formula. In the 1999 constitution, the ND was to get 13% revenue based on the principle of derivation. More states have also been created in the Niger Delta area just like other parts of Nigeria. The states of the ND get more money from the Federation Account than most other states. But what has that got to do with the conflicts in the area? These questions are pertinent since none of these interventions seem to have reduced the restiveness in the ND. What will then be done to reduce the restiveness, especially among youths in the ND?

After enumerating these, the next question is to what extent can we say that the above are conflict resolution initiatives targeted at transforming the conflicts in the ND? Like I said in the literature review, most of these initiatives have been based on political expediency and not on resolving and transforming conflicts in the ND. For instance the creation of twelve states in 1969 by Gowon was principally to break the backbone of Biafra and not to resolve the conflicts in the ND. The tinkering with the revenue allocation formula has also been for the sake of consolidating the various governments' hold on power. As Odia Ofeimun put it, "it may be stated without fear of contradiction that the discovery of oil in the Niger Delta and the rise of oil to the position of principal foreign exchange earner has had more to do with the stakes and whiffs of centralism that have found their way into the constitutions"[10].

At several instances when opportunities for resolving conflicts in the ND have presented themselves, the government has shown a shocking lack of understanding of the situation. A case in point was the Umuechem uprising in 1989. When the people of Umuechem demonstrated against neglect and marginalization, the response was to send police to maintain order. The outcome of that very error of judgment is still haunting the government and in fact the entire Nigeria till today. Second, when after the massacre the government responded by setting up the Justice Inko Tariah panel, it was still obvious that the government was just playing to the gallery of public opinion and not really interested in resolving the conflict. The panel recommended the payment of compensation to those that lost their homes, properties and loved ones. When the checks for the compensation were eventually issued, they bounced at the bank. And almost fourteen years after, no effort has been made by the government to redeem these checks.

The Ogoni case is too well-known to warrant recounting. Odi got the same treatment. When individuals are involved, they either end up in police and prison detention centres or are killed out rightly.

The response of the various state and local governments is to say the least shocking. It is easy to assume that the states and the federal governments are in tandem in their responses to conflicts in the ND. But this may not be altogether so. The state governments in the ND have perfected the act of passing the buck and inciting the people against everyone else but themselves. In July 2002, I was in a delegation that visited one of the governors in the ND. He said that the problem in the ND was the litany of failed promises on the part of the oil companies. As a member of the delegation I wanted to ask him what he intended to do as the governor to ensure that the oil companies fulfilled their promises to the communities? But I was bared by his protocol details.

At other times they make the federal government scapegoat of the conflicts in the ND. A good instance is when they blame the federal government for not releasing the derivation and ecological fund on time[11]. But a survey of the ND states does not show any evidence of the use of such funds when they are released. Already we have made reference to the issue of dud checks issued to Umuechem people. This was by the state government. Under the military, the state government was also used to frame, try and convict Ken Saro Wiwa and eight others.

For the local governments what one can say is that they do not exist. A young man who I trained in conflict resolution described a nearby oil site as their local government. He told me that each day they would go there to see what they would get. I visited one local government in 1999 and was informed, "they were all gone". When asked where they have all gone, a security man on duty said that they had shared the revenue allocation accruing to local governments from the Federation Account for that month and had gone home to come back when the next allocation was due.

By the way this is not peculiar to local government councils in the ND. It is a national malaise. But one would have thought that with the level of consciousness in the ND, the level of suffering and deprivation, and the collective will of the people to redress the wrong done to them over the years, that their leaders would sit up and deliver to their people whatever they could from the little they get. I say this because the ND is in a peculiar circumstance. And apart from probably the Yorubas of the southwest, no other group has been mobilized as the people of the ND.

The worst thing in the response of especially the local and state governments is their inability to provide leadership, articulate the demands of their people and mediate between the communities and the oil companies. This has been one of the greatest sources of instability in the area. The various legislators from the area have not been of much help either. In the National Assembly, most of the legislators appeared to have responded to the conflicts in the ND by simply politicking. By this I mean that they simply say the right things and do nothing. For the various state legislators, their responses have been as inept and lacklustre as the assemblies themselves. Government's response to conflicts in the ND has shown very little creativity and originality of thought.

Oil Company Response

Having seen government response to conflicts in the ND, let us now examine the responses of the oil companies to the conflicts in the ND. The first thing that one notices is that the oil companies were caught unawares. They probably never thought that things would come to a head. But as Ken Saro Wiwa's father put it, if the oil companies were "competent company" they would have been a little more proactive. In general oil companies' response to conflicts in the ND has been twofold; first resorting to the security agencies and second is what I will like to call development dumping.

To my mind there are some possible explanations for this attitude. First, the oil companies have not operated under an enduring democratic tradition in Nigeria. For instance, the colonial government, which appropriated all oil in Nigeria, promulgated the Mineral Ordinance in 1916. By 1937 they had another law, which gave Shell exclusive right to exploit oil minerals in Nigeria. By 1969, the Nigerian government literarily renewed the law with the Petroleum Decree of 1969, which gave the Federal Government exclusive control of oil revenues. So oil company operations have been protected by what we call 'federal might.' This is why they always resort to the use of security agencies. It would also be noticed that out of 44 years of independence Nigeria has only experienced 13 years of democracy. This may not qualify as democracy; civilian rule seems a more appropriate appellation.

There are many explanations for development dumping, one of which is the contract mentality of the Nigerian economy. Development projects afford oil company top management staff the opportunity to award contracts and take 10% kickbacks. It is also an avenue for patronizing traditional rulers and community leaders. Evidence from the various project inspection and evaluation reports show that many projects are either abandoned, in disuse or unsuccessful. I was a member of a delegation that inspected oil company community development projects in 2001. At one of the sites we saw a palm oil mill overgrown with weeds. When I turned on the tap water that was to service the oil mill, there was no water. When I asked one of the young men hanging around the mill why it was not functioning, our guide, an oil company staff, quickly intervened and prevented the young man from responding to my queries. In fact, on our way to the oil palm mill, the community development officer could not find his way to the mill. There were several of such incidents.

The oil companies have also come to see community development projects as an image thing. In a report to Shell on community development, A.R. Melville wrote in 1965:

> '...in the course of my discussion with Mr. Max Davies of Shell/BP Petroleum Development Co. of Nigeria Ltd, the view was strongly put that when it became apparent that oil developments were bringing prosperity to the relatively few in urban areas there might well be considerable resentment in agricultural areas especially where oil operations were involved. What was needed was some sort of public relations effort."[12]

It is easier to point to a school or hospital or a road project than to talk of peace when the military can give you peace of the graveyard. Moreover, oil companies also see community development as charity. They feel that these community developments projects can, and do really assist the people.

But in my opinion, the greatest explanation for community development projects is that they are tacit recognition of the failure of governance in Nigeria. Evidence on the ground shows that many communities in the ND are neglected and suffer from acute lack of basic infrastructure[13]. The oil companies at times do feel genuinely that by providing these amenities that they are filling in for the inadequacies of government. In fact many people in the ND see the oil companies as alternative local government. This has its own implications for the conflicts in the ND.

Whenever conflicts arise between the communities and the oil companies, the immediate response of oil companies has been to invite security agencies. Umuechem is a good example where Mr. Udofia, a Shell staff specifically asked for Mobile police. The militarization of the ND is all part of this response.

Human Rights Watch and Amnesty International have documented some of the fallouts from this reflex resort to military solution. There are many reported cases of extra-judicial killings; entire communities are sacked, and there are incessant killings. The situation is so bad that top officials of oil companies go about with military aides. In July 2003, when I joined an intervention team to work on an oil spill in the western part of the ND, the managing director of the oil company left the venue of the meeting with a truck filled with armed mobile policemen who escorted him to the airport[14].

As part of this security arrangements, oil companies award what they call surveillance contracts. Whenever an oil company or their contractor has an issue with a community, part of their response is to offer the youths of the community money to protect their oil installations in the area. This arrangement has bred its own conflict, which right now is not within the purview of this book.

But suffice to mention that the jockeying for these surveillance contracts and the subsequent sharing of the proceeds from these contracts constitute a potential source of explosive conflicts. For instance in one community in eastern ND the traditional ruler complained of how an oil company terminated the surveillance contract because according to him "we are a peaceful community". He asked whether the oil company wanted them to resort to violence before they would renew the contract. This is a zero sum game for the oil companies. If they award the surveillance contracts it will lead to conflicts, if they do not, it will lead to the vandalisation of their equipments. Either way they lose. What is the way out?

Apart from the above, the oil companies also do what I will refer to as development dumping. This is done by simply dumping a development project in the community. For instance after the Umuechem uprising,

Shell went in and began to build a hospital. Midway this project was abandoned. A journey round the ND will show so many of such projects in the ND. In fact the projects are so many that yearly Shell invites what they call stakeholders to audit their community development projects. These projects are usually developed in communities that are restive. And of course there have been development projects to reward the so-called peaceful communities. [15]

Needless to say that these projects have not reduced the number/incidence of violent conflicts in the area. Rather it has worsened every passing day. The question is why is it that in spite of the huge amount of money spent on community development projects in the ND, the area remains largely restive? Part of the answer lies in redefining conflict resolution initiatives in the ND.

As part of their response to conflicts in their area of operations, oil companies also award scholarships and offer employments. The oil companies determine the criteria for the award of the scholarships. They also determine how many should be awarded. The question is this, how will the oil companies administer developments projects in such a way as to "do no harm"?[16]

At Umuechem the people complained that the number of scholarship forms given to them by Shell was too small. Those who completed the forms qualified to sit for the scholarship test. This was when I was carrying out the intervention at Umuechem. We decided to approach Shell to ask for more forms. But before that I asked the people what the criteria were for issuing the scholarship forms. They said they did not know, that it was Shell's decision... We went to Shell to meet with the man in charge of the area. When I asked him the same question of the criteria used to issue scholarship forms, he told me the number of barrels of oil produced from the wells in the community. I asked him again who the custodian of this figure was. He replied, Shell. And I said, "considering your experience with the people do you think that they can trust you to give them the right figures?" In reply, he opened his drawer and brought out two more forms. I wondered to myself if community development could be this arbitrary and haphazard?

Oil companies also offer employment as part of their response to conflicts in the ND. It is obvious how convoluted this issue of employment can be. In an agreement between Eleme and Okrika people over employment at the Eleme Refinery in Port Harcourt, Eleme people were given 60% of employment opportunities in the refinery; Okrika people were given 30%, while the rest of Nigeria had 10%. Needless to say that this has not ended the conflict in the area, rather it has escalated it. In another instance, the Egi Women Movement in their charter of demand asked for the following "an 80% quota for Egi indigenes including nurses at the lower cadre levels (01-06) in Elf and all Elf servicing firms working in the area."[17] Continuing they asked for "a 60% quota for Egi indigenes at the middle level manpower (07 and above) in the Elf and Elf servicing firms working in the area."[18]

A careful reading of the quoted sections of the charter of demand however shows that it is inherently conflictual. For instance, how can they monitor compliance? Two, do they have the staff to fill up these vacancies? Three, would the companies have vacancies in these areas? Four, would the indigenes of Egi pass the interviews? These are some of the issues that need to be addressed in this kind of response.

Community Response

Meanwhile, before redefining conflict resolution initiatives in the ND, let us examine how communities have responded to conflicts in the ND. For most communities the first thing they do whenever they feel that they have a problem with an oil company is to write them a letter[19]. They claim often that the oil companies neither acknowledge receipt nor respond to their letters. I was informed that they write these letters to seek audience with staff of oil companies to air their grievance and begin a process of dialogue. When these fail they employ other means to attract attention.

At Umuechem they demonstrated, in other areas they may block access to company facilities; they may even besiege the company premises as the women of Ugborodo[20] did. In extreme cases they may take oil company staff hostage as they did in Sangana local government area of Bayelsa State[21]. According to Oronto Douglas, "we do all these to draw attention to our plight"[22].

To my mind the most comprehensive response so far has been the establishment of MOSOP. This is important because for the very first time the political establishment in Nigeria panicked. The establishment panicked because MOSOP internationalized the issue in the ND and made extensive use of the media. Second, the agitation was anchored around such issues as human rights, environmental integrity and self-determination. All these resonated with the international community especially human rights and environmental NGOs.

With the end of the Cold War, NGOs began to play a critical role in world affairs. The ND issue latched on this new phenomenon. Again the question is this, was the establishment of MOSOP a response to the conflict in the ND or a mechanism for its escalation?[23] Kriesberg[24] says that it is escalation. He argues that when conflicting parties establish specialized agencies to respond to conflicts that it is an escalatory mechanism and not resolution.

Before the establishment of MOSOP the people of Ogoni were suffering. I do not think that MOSOP's aim was to escalate the conflict. Rather I think that the main reason for the establishment of MOSOP was to comprehensively articulate the aspirations of the people of the ND in a coherent manner. Second, was for the people to have a credible voice with which to negotiate.

Third is that there have been conflicting signals emanating from the ND. The establishment of MOSOP was to standardize their demands. Four, a tradition of working through associations is embedded in the culture of the people. So this was a reinforcement of that culture. Oil companies, government and NGOs also prefer to work with groups. It is possible that this could account for the establishment of MOSOP and others. But more importantly was that MOSOP came into being in the heat of the much-vaunted IBB's truncated transition programme. It could also be that Ogoni people wanted a common front to collectively articulate their political destiny.

MOSOP and others did not just blow hot air. They concretized their agitation with striking symbols. For instance there was the Ogoni Bill of Rights, Ogoni National Anthem and Ogoni Flag. They followed this up by declaring Shell *persona non grata* in Ogoni land. MOSOP sought recognition with the UN and so began the greatest mobilization ever known in the history of the ND.

Having opened the way, so many ND communities followed in the Ogoni footstep. First, in 1992 the Movement for the Survival of Izon Ethnic Nationality was established. It also developed the Izon Peoples Charter, which drew from the Ogoni Bill of Rights. In 1997 the Chikoko Movement,[25] an anti-oil spillage pressure group, was launched at Ogbia. There was also the awakening of such groups as Egbesu Youths, Isoko Ethnic Minority Rights and Environmental Protection Council, and the Ijaw Youths Council. On December 11, 1998 the Ijaw Youths Council adopted the Kaiama

Declaration. Ever since, there has been a rapid escalation of conflicts in the ND. The paradox of the ND conflict is that as the government try to repress the people of the ND, the more the conflicts in the area seem to escalate.

There are many possible explanations for this. One very important explanation is that the conflicts are being suppressed not resolved or transformed. There is the fear that if something drastic is not done to arrest the situation, it might probably explode. For instance, some groups in the ND are now calling for outright secession. More in depth and theoretical explanations shall be offered for the escalation in subsequent chapters.

Response of Non-governmental organizations (NGO)

The next stage of our work will be to examine the responses of the various non-governmental organizations to the conflicts in the ND. Earlier on in this chapter I have also alluded to the growing profile of NGOs since the end of the Cold War. With the reputation of Ken Saro Wiwa as a writer and the mobilization of his people, the ND was put in the spotlight. When Ken Saro Wiwa and eight others were executed in 1995, the outrage generated by that judicial murder drew the attention of the world to the ND. This explains why the ND became a kind of berthing ground for NGOs. Apart from the national and international NGOs that intervened, there was also the sprouting of local ones as we have identified before[26].

On arrival the first thing that the NGOs did was to publicize the issues in the ND[27]. They published reports and posted same on the web. For instance for this very project I have seen, and read close to one hundred reports on the ND alone. These reports began to reverberate in major centres of the world. One of the major outcomes of these reports was the awareness it created about the goings on in the ND. In 1997, the

Ecumenical Council on Corporate Responsibility asked Rev. Fr. Kevin O'Hara (an Irish missionary priest and human rights activist) to undertake an inspection of oil companies' development projects in the ND. When Fr. O'Hara produced his findings, a few heads rolled in one of the oil companies.

This started a round of visitations[28] by these NGOs to see things for themselves. This has two major impacts. First, the ND people began to realize that they are not alone, that their plight is known worldwide. Second, it opened the doors of European and American countries for ND people who felt threatened and wished to seek for asylum. Many took advantage of this and fled. On getting to these countries, they studied and began to mobilize resources and campaign all over Europe. They also began to raise money for fellow activists who could not make it abroad.

After the death of Abacha in 1998, many began to return home. They were now joined by their friends from abroad who wanted to see things for themselves. I have hosted more than five of these visits. And at the end of every visit, the visitors went back with the actual picture of what was going on in the ND, and they began to campaign for, and on behalf of the people. Many also issued report of what they saw[29]. In fact some also produced documentaries on the issues of the ND. One of the most striking of these documentaries is Delta Force aired on Channel 4 in Ireland.

Some started training programmes. The aim of these training programs was two-pronged. First was to build the capacity of the people of the ND to address their own conflicts. In this area, a pioneer NGO is

Academic Associates Peaceworks (AAPW). This NGO conducted several trainings in the ND. Many of their trainings were during the peak of the Eleme/Okrika conflict in the ND. AAPW also did extensive work in the Warri area. The outcome of that intervention in Warri is the book entitled "Conflict and Instability in the Niger Delta: The Warri Case"[30]. The Constitutional Rights Project (CRP) also conducted several trainings in the ND. The Committee for the Defence of Human Rights (CDHR) has also been very active in training programs. The various Justice, Development and Peace Commission (JDPC) of the Catholic Church have also conducted several conflict resolution trainings in the ND.

I have been very active in most of the training sessions. One of the major contributions of these trainings is that they give you an insight into how the various parties view the conflicts in their area. For instance, it was in one of such trainings that I found out that the whole idea of self-determination is seen by the people of the ND within the context of the evolution of Nigeria as a nation. For instance the people of the ND at that training explained that when other regions had their natural resources like cocoa from the western region, palm produce from the east and groundnut from the west, the various regions controlled the production and revenue from these resources. But in the case of the ND that is not the same.

Many other issues also come to the fore during the trainings. For instance, the impression has been created that the people of the ND want their oil resources alone. But in these trainings one will observe that this is just a position, not an interest or need.

Many NGOs have also carried out several interventions in the ND. They have really assisted the communities in mediating their conflicts with the oil companies. One that readily comes to mind is Community Rights Initiative[31]. This NGO was instrumental in mediating the conflicts between the Nigeria National Petroleum Corporation and the Egi community.

I have been very fascinated with the work of the Centre for Social and Corporate Responsibility (CSCR). The CSCR was founded by Rev. Fr. Kevin O'Hara, an Irish missionary priest and human rights activist and has operated from outside the ND for sometime. But in 2001, they opened an office in Port Harcourt. This group started as a human rights group. But in the course of their work, they found out that there is more to the ND issue than mere human rights violations. For instance most issues of human rights violations had their roots in conflicts from the various communities. For instance it was not uncommon to find police detain young men for armed robbery when they were suspected to have vandalized oil company installations. With this realization, the CSCR revised its strategy of operation.

They called their new strategy, "shareholders' leverage". The analysis of this strategy and its conceptual framework and application will be in the subsequent chapters.

Meanwhile what are the lessons or insights from the above analysis of the responses? The most important lesson is that almost all the parties have responded to the conflicts in the ND in similar manner. For instance, the governments have used the courts. The government of Obasanjo took the ND states to court to stop the on/offshore oil dichotomy. The government even followed it up with a bill at the National Assembly.

The oil companies have also been in and out of courts with the communities for a very long time. For instance since 1990 the people of Umuechem have been in court with Shell over their issue. The Ogoni are also in court over oil spills. These court cases are protracted and last almost a lifetime. Some NGOs have also gone to court on behalf of some people in the ND in what they refer to as strategic impact litigation or public impact litigation[32].

All these are meant to draw attention to issues in the ND. But unfortunately, none of this has ever produced the desired result. Rather they have ended up complicating the issues at stake.

In my opinion, the most important pitfall of the various efforts to resolve the issues is that there has been no coordinated attempt by all the parties to constructively engage the conflicts in the ND. This is because most of the parties have not seen what is happening in the ND as conflicts but as part of the evolution of the Nigerian nation. Ihonvbere argues that "the crisis in the oil producing communities is only symptomatic of the crisis of the Nigerian state."[33] Second, the politicization and internationalization of conflicts in the ND have given the conflicts a life of their own. Third, there seem to be no credible party that will consistently engage and intervene in the conflicts in the ND. The governments that would have filled this gap unfortunately are parties to the conflict.

Finally, the tying of the conflicts in the ND to the issues of slavery, colonialism and racism seem to becloud the issue and overwhelm potential interveners. But the irony is that those who fund the intervention work in the ND are also accused of being behind the oil companies at whose doorsteps the people of the ND have dumped their frustrations. These are some of the challenges that we shall face in the next chapter.

One of the uniqueness of the responses is that they have all been in the form of a demand for something. Second, none of the parties seem to have been able to gain a foothold into the other party's table. Each party works at its own pace, in its own manner and at its own time. There are also structural impediments to responding in a certain manner. For instance due to lack of access to justice and the courts, communities and individuals tend to be unable to sue the companies. Due to the same constraints, it is also almost impossible to sue such organizations as the NNPC. Therefore in order to bring them to the negotiating table, the communities use all kinds of tactics.

However, the CSCR has been able to gain a foothold inside some oil company offices especially Shell. Though this has given them some kind of leverage in their work, in a publication recently, the CSCR lamented its inability to really make appreciable progress. In their assessment of five case studies[34] namely Umuechem, Ojobo, Ogbodo, Batan and Oloibiri, the CSCR did not have any cause to cheer their partnering with Shell based on their model of "shareholder's leverage" (SL). Though it is too early to carry out a comprehensive assessment of the use of (SL), it appears to have a lot of promise because embedded in it is a mechanism for collaboration.

Notes and References

[1] Frynas, J.G. Oil in Nigeria. P.226 and 229.

[2] Lederach, J.P. Building Peace

[3] Curle, A. Tools for conflict transformation

[4] Frynas in Oil in Nigeria, identified many obstacles to access to the courts in general.

[5] Saro Wiwa, Kenule. On a Darkling Plain. London: Epsom, 1989, p.199.

[6] See Osaghae, E.E. "Ogoni Uprising". P.337.

[7] See "Boiling Point." P. 51.

[8] Between 1999 and 2002 I visited Ogoni more than five times. During this period the importance of these labeling was brought home to me. For instance in 2002 July a young man in his teens has warned me not to talk to a certain chief for he is a vulture.

[9] See Punch Newspapers, Friday, August 6, 2004.

[10] ERA/FoEN: The Emperor Has No Clothes. Benin City: ERA/FoEN, 2000, p.65.

[11] See Thisday August 6, 2004.

[12] Quoted in Frynas, J.G. Oil in Nigeria, p.51.

[13] I have conducted several interventions in the ND and have also visited many communities in the ND. There is clearly a lack of basic amenities in the area

[14] The details of this intervention have not yet been made public since we are still working on it.

[15] See Community Development a quarterly publication of Shell, Vol.1, No.1, July 2001.

[16] See Mary B. Anderson. Do No Harm (www.cdainc.com).

[17] Egi Women Movement: Charter of Demand, article 2, section 1.1 (1999)

[18] Ibid: section 1.2.

[19] I used this as a role play in a workshop, which I facilitated for youths from the ND at the International Airport Hotel at Omagwa Port Harcourt. In that workshop I have asked a young man why they respond the way they do if oil companies do not respond to their letters when they do not respond the same way when government agencies do not reply their letters. The young man's response was that I am an oil company agent.

[20] See The Guardian, Sunday, August 25, 2002.

[21] See Thisday Newspaper, July 19, 2004.

[22] Interview with Oronto Douglas at ERA/FoEN office in PH, August 2002.

[23] For details of this model of intervention known as ARIA, see Rothman, J. Resolving Identity-based Conflicts.

[24] Kriesberg, L. Constructive Conflicts. P. 176.

---

[25] The Chikoko Movement has been accused of being a terrorist organization that preaches violence, in an interview for this project, Oronto Douglas denied this allegation.

[26] See Ikelegbe, A. The Journal of Modern African Studies, 39, 3 (2001), pp.437-469.

[27] One of the most comprehensive reports on the ND is The Price of Oil by Human Rights Watch (1999). See also Land, Oil and Human Rights in Nigeria's Delta Region by Constitutional Rights Project (1999).

[28] Several NGOs like Trocaire, Oxfam, Cordaid, Misereor etc. sent delegations to do an on the spot assessment of the situation in the ND. Between 1999 and 2002, I participated in at least five of these visitations.

[29] See when the Pressure Drops By ECCR for instance and the report by Corporate Engagement Project.

[30] Imobighe, T.A. et al. Conflict and Instability in the Niger Delta: The Warri Case. Ibadan: Spectrum Books, 2002.

[31] I interviewed the founder of this NGO Mr. Wisdom who told me this and gave me documents to back up their work.

[32] Frynas, J.G. Oil in Nigeria. P.90 reported the Douglas v. Shell case Suit no. FHC/L/CS/573/96 in the Federal High Court, Lagos.

[33] See Boiling Point. P.74.

[34] See Extractive Industries and Corporate Responsibility: Case Study of CSCR (2004) by Drs E. O. Emmanuel and David C. Okwudili.

Chapter 6

Redefining Conflict Resolution Initiatives in the Niger Delta

In this chapter I am going to do a redefinition of conflicts in the Niger Delta (ND). This redefinition will not be based on the rationalist approach of absolute truth. It will be based on my experience working with the people. I will also draw on my experience as someone from the ND. And I am going to draw from my academic training in conflict transformation and conflict analysis and resolution. Finally, I will be sharing the findings from my two years of fieldwork research on the issue.

This process of redefinition is going to be anchored around the ten propositions of Byrne[1], which he distilled from transformational conflict resolution literature. I am using these propositions because they encapsulate the core principles and values for building sustainable peace. The first proposition is that "reconciliation is a multifaceted idea built on truth, mercy, justice and peace."[2] This first principle is very important because without reconciliation, parties to the conflicts in the ND cannot purge themselves of hate, anger and fear which has dominated and consumed them throughout the period of the conflict. The issue then is how do we foster reconciliation among the parties in the conflicts in the ND? Second, when we foster this reconciliation, how do we sustain it?

The second proposition is what Lederach refers to as indigenous empowerment. This process of indigenous empowerment enables both the parties to the conflict and the interveners to discover what David Cooperider calls the life-giving force of the community. With the barrage of activities taking place in the ND, the people feel totally lost as to how to be part of what is going on. This is why they tend to see everyone that comes into the community as a potential messiah. But if they are empowered their collective esteem will be restored and they will be in a better position to engage the issues that confront them. For instance, the ND has one of the best collections of talents in the whole of Nigeria but when it comes to issues that concern their communities, they are often powerless. This process of indigenous empowerment might be the answer to restoring their collective dignity.

The third principle is that transformative politics must mainstream and promote participation of all in the democratic process. This helps to build confidence between the parties. And in a situation like the ND where the people have been conditioned to believe that their woes are as a result of their being minorities, this principle is very useful. Elsewhere in this book I have made the point that a key problem, which I noticed in the ND, is the absolute non-recognition of democratic structures. In my opinion, this is not unconnected with this minority complex, which has led them to believe that as minorities they do not count since democracy is a game of numbers. How do we then encourage them to participate in the democratic process? It will be germane to point out that they are not the only ones in Nigeria that have been apathetic to democracy. Theirs however seem to benumb.

The fourth proposition is that there should be a link between personal participation and political empowerment. This is true of the conflicts in the ND. The youths, men and women in the ND are very articulate and conversant with the issues in their area. This is how Thelma Ekiyor reported this awareness among the youths of the ND:

"…exploitation, marginalization and oppression, I heard these [sic] at all the workshops we organized. Even the young boys, who do not fully comprehend the meaning of the words cry them."[3]

The issue then is how can this awareness of the issues be harnessed to cooperative relationships, trust and confidence? It is my belief that by tapping into this conscientisation, the people will be empowered, their self-esteem will also be restored and they will be encouraged to participate in any peace process that is put in place.

The fifth principle, which Byrne mentions, is that of using peaceful means to protect the rights of minorities. The implication of this for the ND is that there will no longer be any need for their grievances to be ignored. Everyone will be aware that they are a minority and their rights will be protected, not in the form of tokenism but as genuine and respected members of the polity.

This will include recognition that the oil wealth is located within their area and that they bear the brunt of the hazards of its exploitation. This may also include some form of constitutional guarantees. Some of the mechanisms of the above have already been provided for in the Nigerian constitution. Such issues as federal character, fundamental human rights, and revenue sharing have all been provided for. The issue is that of the mechanism for implementation.

Reconciliation, recognition of differences and inclusiveness involve a collective visioning of the future. For instance Ekiyor has described the mindset of the people of the ND over their plight thus:

"…they believe that if this environmental degradation is allowed to continue, their source of livelihood will be destroyed forever. They also believe that if the oil and gas wells run dry in the

Niger Delta, the multinationals will move out to other prospects, the government will look for another source of its revenue and the people of the Niger Delta will be left with desolate land and polluted air and water." [4]

To jointly design a future where all will be free from fear and hatred is part of the transformation process. In the ND today, this prospect is non-existent.

Transforming relationships especially in divided societies is a collective enterprise since everyone has a stake in peaceful co-existence. Lederach makes the case of involving the three levels of actors. This is because if all are involved, the people will own the process and this will build trust and confidence. Right now in the ND different level of actors are working at cross-purposes. This is made worse by the fact that each level of actors feels that the other level is working against its interest. For instance I have illustrated the conflicts between the youths and the elders. I have also discussed the no love lost relationship between the oil companies and NGOs on the one hand, and between the government and NGOs on the other.

And if we must build sustainable peace in the ND we must bear in mind that development is a holistic concept. For any developmental effort to make sense it must incorporate and promote change and growth both at individual and community level. And to have development, which will be far-reaching, we must mobilize the social and cultural resources of the people. This process of mobilization can only take place in an atmosphere of respect, trust and peace. As a principle of conflict transformation, development must promote growth and change.

Before one can transform structures, institutions and relationships, the individual must first of all be transformed. This process of spiritual transformation is such that one must believe in the values of peace to work for peace.

For an individual to work for truth and justice in a non-violent manner, the individual must believe in those ideals. Not only believe but must be committed to them. One of the crises plaguing the field of conflict transformation is that it has become another academic endeavour without a soul. People work into workshops or conferences and extol all the values of non-violence but often unwittingly believe in violence. Even activists sometimes actively encourage the use of violence but usually deny doing so publicly. So this process of transformation must start from the self.

As I have noted elsewhere in this book, people construct conflict from their perception and interpretation of events. The process of transformation must involve a re-conceptualization (call it deconstruction if you wish) of not only the conflicts but also the events, parties and context of the conflict. Each party's perception of conflicts is like a burden, which they carry about. And like a heavy burden, it weighs them down in the process of transformation or as Lederach puts it in the journey towards reconciliation.

The above are some of the ways that adversarial relationships are transformed and this fosters reconciliation. These propositions as we have seen incorporate such values as inclusiveness, participation, non-violence, respect and recognition of differences and cooperation. The above review shows us that most of these values identified as pathways to conflict transformation and peacebuilding is almost non-existent in the ND. Bearing all the above in mind let us now redefine the conflicts in the ND as we saw them with a view to suggesting methods and mechanisms for conflict transformation and peace building in the area.

My findings show that we can no longer speak of the conflict in the ND. Initially, the impression given was that there was only one conflict in the ND that was replicated in various communities. Many writers have equated different causes with different conflicts. What is apparent is that the eight different types of conflicts identified in the ND are not the causes of the conflicts.

The implication of this "hierarchical structure of the conflicts" is that we can now know exactly what specific and relational issues we are addressing. More importantly, it helps us in knowing where to exactly focus when we engage in the transformation of individuals, institutions and laws.

Another interesting finding of my research is that there are communities that are not responding to these conflicts like others. For instance in eastern ND like Asa, and Ndoki, their responses to the conflicts were not the same with what I saw in western ND. Nevertheless, they are still aggrieved and bitter at the oil companies and government. This means that in designing an intervention plan we must be very careful in identifying the critical and strategic needs of the communities. A holistic approach will give us an overall picture of the situation. This is important because a piece-meal approach that is not integrated will leave gaps for the explosion of more violent conflicts.

Byrne's social cubism[5] will be a very useful lens for a more holistic view of this redefinition. Byrne and Carter identified six forces "that combine to produce patterns of intergroup behaviour." [6] These forces are historical, religious, demographic, political, economic and psycho-cultural factors. One important thing to note about these factors as used by Byrne and Carter is not that they cause conflict or produce adequate explanations for conflicts but that they produce patterns of behaviour for groups especially when they relate to each other.

Let me illustrate some of these forces. In the ND for instance, almost all the issues about their relationship with the oil companies are embedded in what the "whiteman" did to their ancestors. The import of this is that a deep-seated feeling of resentment is already inside the average ND indigene. This is passed on from generation to generation. So when the average ND indigene sees an oil company staff today, they do not see an employee but "a slave dealer," "a colonial official" and "a dubious missionary". This explains why it will be difficult to build trust between the host communities and the oil companies because these images carry a lot of meaning in the history of the evolution of the ND.

Let me give another example. In their relationship with the other ethnic groups in Nigeria, the people of ND feel that they are a minority. In fact they believe that the treatment that they are getting is because they are a minority. This conceptualization of themselves as minorities is what Byrne and Carter refer to as demographic force. In the case of the ND there have been many actions of the Nigerian state that tend to reinforce this perception of being maltreated because of being a minority. Again going through the history of Nigeria one notices a consistent pattern of discrimination against minority groups. This has bred mistrust and hate. In intervening in the conflicts in the ND therefore, these are not issues that will be glossed over.

The interesting thing about these forces is that it is a combination of what Byrne and Carter refer to as "objective and subjective" perspectives. As Byrne and Keashly put it:

> "...merging these two perspectives focuses attention on recognizing that protracted inter-communal conflicts develop and are sustained through the complex interaction of a number of factors......consequently, a multiplicity of intervention efforts must run simultaneously and sometimes sequentially at all levels to de-escalate the conflict through all of the stages of escalation and to transform the underlying nature of conflict over time."[7]

The next issue to confront is how do we conduct an intervention that will take into consideration all the above forces. Lederach recommends an integrated approach to peace-building, which we have discussed earlier. But Lederach emphasizes reconciliation, while Byrne and Keashly recommend a multiplicity of intervention efforts that must run sequentially and simultaneously at all levels. But before this could be done, we need to de-escalate the conflict at least within the crisis and short-term periods. For us to be able to do this, we need to understand why the conflicts in the ND seem to be escalating. Since there is no evidence to suggest that there is a higher number of deaths, hijacking or hostage-taking in the ND in any given period, the views expressed here will be merely speculative, and at best a suggestion.

The main issues to address here is why, and how conflicts in the ND have escalated. By escalation I do not necessarily imply more deaths or increased level of violence but I mean the intensity of the division and polarization of the ND. This distinction is important because there may be a high number of deaths in a particular uprising like Odi while there may be more uprising with a lower death toll. For instance, a report by Environmental Rights Action puts the death toll at Odi at 2,483[8]. For any quantitative study, this figure will skew the findings.

The division and polarization in the ND has been deepened because as their cause is more valued and draws sympathizers, acts of violence are justified by referring to the other camp. In the ND, this applies both to the oil companies and government. The oil companies are making increased use of security; their argument is that the people of the ND are becoming more violent. The government is using more repressive measures and the justification is that there is a breakdown of law and order. The communities are becoming more incensed because according to them the government has become more repressive while the oil companies are more careless. This is very close to the concept of security dilemma in international conflict.

Closely related to the above is a sense of heightened commitment to the cause, and the conviction that the goal is attainable. The conflicts in the ND have become a kind of self-fulfilling mission for those involved in it. Major Okuntimo[9] may probably have gone through the Nigerian Army without anyone hearing of him. Today he is a part of history, (though many may argue on the wrong side of history). Many young people today in the ND have no career or achievement to their name except for being involved in these conflicts. Still many others are making careers in the various oil companies because of these conflicts. In fact Governor James Ibori of Delta State appointed Ovuozorie Macaulay as Commissioner for Inter-Ethnic Relations and Conflict Resolution[10]. All these point to the fact that there is a renewed commitment to the whole ND issue and a belief that "victory is certain".

Apart from this belief that victory is certain, there is also the tendency on the part of the parties to increase their demands as they get more concessions. At this point other parties will resist this and the fall out is a stalemate, which inevitably escalates the conflicts.

As commitment grows and conviction increases so also the influence of leaders who are ever ready to escalate the conflicts. The conflicts in the ND have thrown up people of different character in government, the oil companies, the communities and NGOs. These leaders have become so influential that their followers can follow them to battle blindfolded. Kriesberg refers to these people as hardliners. In the ND some of these leaders are hardliners while others are moderates. The critical factor is that they have a mass of people at their beck and call to do their bidding daily. How the influence of these leaders can be harnessed for peaceful purposes is part of the multiple intervention efforts referred to by Byrne and Carter.

Kriesberg[11] refers to what he terms the concept of entrapment. This concept states that the more groups commit resources to a conflict, the more they are reluctant to get out of it because that will mean losing all that they have already committed. This can be seen on all sides of the conflicts in the ND. Some of the brightest and best of the ND people have been lost to these conflicts. So much resources and energy have been committed to the conflict; a withdrawal will mean to acquiesce to the 'oppressors'.

As the issues involved in the ND conflicts expand and other parties get involved, there is a polarization of relationships, and this eventually leads to selective perception of the issues involved. And the involvement of many parties over many issues complicates the dynamics of the conflicts in the ND. Recall the eight conflicts identified and some of the issues raised when we reviewed the literature of conflicts in the ND. All these are issues that need to be addressed in designing a comprehensive intervention plan. Apart from this multiplicity of issues, there is this dimension of superimposing a conflict on another one. Frynas[12] gave a good example with a land case where an oil company complicated the situation by paying compensation to one of the parties. Moreover as the number of issues in a given conflict increases so also the anger, hate and polarization that follows. All these we saw in the conflicts in the ND as we discussed them in the previous chapters.

The conflicts in the ND have also been reframed into what I will refer to as grand ideological issues. This has enabled each party in the conflict to attract more sympathizers. For instance, it will sound mundane, almost pedestrian, to argue that one is taking an oil company staff hostage for a land dispute. But it is more rational to say that you are fighting for such grand and universal values as minority rights, environmental integrity, true federalism, political marginalization and democratization. These are issues and concepts that resonate with people all over the world. They are also relatively easy to articulate. Moreover the world through such organization as the UN has developed instruments for addressing these issues.

These grand concepts encapsulate collective goals and aspirations. The universality of these values legitimises the struggle and involvement in the conflicts. It also justifies the means through which these ideals are fought. These universal ideals become instruments for operationalising the conflict, thereby etching on it the imprints of a survival strategy. It therefore acts as an instrument for mobilizing resources for the conflict.

This has its own unintended consequences. For instance, under the military, the conflicts in the ND were seen as a direct affront on the military. Again with the advent of democracy in 1999, the conflicts were seen as political opposition meant to destabilize an emerging democracy. This therefore obscures the specific and relational issues that needed to be addressed before attending to the systemic issues. This could be seen from the activities of the United States Agency for International Development (USAID).

USAID's intervention in the conflicts in the ND was anchored around the Office of Transitional Initiatives. The assumption within the bureaucracy was that most of the interventions done were to stabilize democracy in Nigeria. It is obvious that democracy instead of de-escalating the conflicts in the ND has paradoxically intensified it and has even reinforced and entrenched the feelings of marginalization of the people of the ND.

As these issues are piled on each other, the parties feel overwhelmed by the burden of injustice and oppression. This is often complicated by the unresponsiveness of the government and the oil companies. The parties therefore come to believe that they are fighting for their lives and are justified to use any means available to preserve themselves. In this instance, the government and even oil companies may employ more repressive measures, which will only escalate the conflict further. This is a cycle that goes round and comes round, leaving in its wake a trail of sorrow, hate, tears and blood.

Another issue in the ND which has polarized the parties is what I would term the "professionalisation of agencies" by the conflicting parties. Today, we have several task forces and agencies of governments whose roles are mainly to respond to the conflicts in the ND. This has led to the creation of extra-judicial bodies and organizations. For instance in the preceding chapters we have made reference to the Joint Task Force on security in the ND. We also talked of the Internal Security Force led by Okutimo.

A very interesting scenario was played out when the women of Ugborodo took over Chevron oil installations. At the end of the siege period, we read that the Federal Government of Nigeria have decided to establish an all female mobile police force. Obviously all these institutions will do all they can to ensure that these conflicts continue because if the conflicts come to an end, then their jobs will also come to an end.

The oil companies also have their own sundry bodies - both open and clandestine, whose main mandate is to deal with these conflicts. The communities have so many organizations whose main motivation is to engage in the conflicts in the area. In fact many youths have decided to make career out of these conflicts.

On one occasion I visited a community while I was intervening in a conflict situation. A young man, who was about 20 years old, blocked the road and said that I should not pass through the road. Having worked in the ND for a long time and coming from the area, I knew how to navigate my way around the ND. I alighted from the vehicle and engaged the young man in a friendly conversation as to why I was in the community. After which we struck some kind of friendship. I invited him to come to the office, because I was genuinely interested in making him part of our intervention team. He told me bluntly that he would not come, that he made more money collecting tolls on the road from passers-by. Till date I have maintained my friendship with the young man. But that is a how specialized individual and agencies have formed around the conflicts in the area.

NGOs are not left out in this professionalisation of the conflicts in the ND. Sam Doe, the executive director of the West African Network for Peacebuilding (WANEP) refers to it as the "peace industry". Apart from the proliferation of NGOs, there is also the proliferation of projects. The implication of this is that people without appropriate skills are all over the ND, sometimes doing more harm than good. In fact many trainers in the name of capacity building in the ND sometimes incite the people after which they leave them to their fate.

There are usually no follow-ups to interventions, no appropriate mechanisms for evaluation of interventions and no mechanisms for sharing best practices. In our redefinition of conflicts in the ND, these are issues that we must address. Earlier on, I have made reference to the case of the bounced checks of the people of Umuechem. If the NGOs in the area recognized their duty to the people, then something could have been done about those checks. Throughout my work and research in the ND it was only the CSCR that tried to engage the government on this issue of bounced checks. It is hard to imagine that the NGOs could not even go to court on the behalf of the people of Umuechem over this issue

Having redefined some of the issues involved in the ND conflicts; let me turn to the issue of redefining the modes of intervention. Earlier I have cast my vote for Lederach's integrated peace-building framework. I need not repeat myself on this issue. But later we are going to draw from Lederach's conception of reconciliation as part of the integrated framework for peace-building.

Second, I am also proposing what I call the Hierarchical Approach to Peacebuilding. And finally I shall draw from Fetherson's[13] theoretical conceptualization of peacekeeping. This will also be in line with his emancipatory conflict transformation. By emancipatory I am not going to use emancipatory in the doctrinaire fashion in which Fetherson employed it. However my use of emancipatory in this context is in line with Lederach's conception of reconciliation, which emancipates parties involved in a conflict from embedded hatred, malice, poverty, disease, anger and fear. This process of emancipation opens them up and frees them from a selective perception of the other.

Fetherson defines peacekeeping as

"...an intervention utilized only after a conflict has become violent and protracted. Its functions, therefore, must be at least two-fold: first, to act as a means of separation, a breathing space where both sides can step back from confrontation; second, and crucially, peacekeeping functions as peacebuilding – working on improving communication and on social, political and economic regeneration."[14]

The contention here is that in the case of peacekeeping at the international level, regional groupings and international organizations like the UN, AU, and ECOWAS have been at the forefront. In the case of the ND who will do this? How can it be accepted that the situation in the ND warrants a UN-type of peacekeeping?

These questions arise because the government of Nigeria has largely failed to live up to expectation as a credible third party in the conflicts in the ND. And we should not draw attention to the ND by making the kind of comparison that Toluragha made with Iraq. This will be playing to the gallery. By peacekeeping here I do not mean a further militarization of the ND. What I am calling for is a coalition of credible individuals and groups that will monitor the situation in the ND and report accurately on what the situation is. These groups are not going to be outsiders. In fact I would recommend what Lederach calls "insider partials".[15] These are individuals and groups with interest in the transformation of the conflict. These groups include such oil company shareholders like the London-based Ecumenical Council for Corporate Responsibility (ECCR), the New York-based Interfaith Centre for Corporate Responsibility (ICCR), the Eminent Persons which the Church World Service was proposing to establish. From the government side we can have the National Assembly committee on Human Rights or any of such committees in the legislature whose mandate includes such issues. The oil-bearing communities would be allowed to nominate their own representatives. This proposition is also based on the concept of superordinate goal, which according to Byrne and Keashly "is a goal that all parties want to achieve but cannot be achieved without the participation of all parties."[16]

I propose these groups because they are made up of people who can cut across the three levels of Lederach's pyramid of actors. They are also credible and could interface with government. Moreover they will be respected both locally and internationally. The key thing in putting together these groups will be to ensure credibility and acceptance by all groups involved. The ECCR and ICCR have over time proven themselves to be very responsible organizations that are concerned about where their money is invested.

Another group of people that will play a credible role in this our conceptualization of peacekeeping are the various unions that exist in the oil companies. A few years back, I conducted training for labour leaders in Nigeria. This training was sponsored by the American Centre for International Labour Solidarity. That training was designed for labour leaders to broaden their conceptualization of leadership from the narrow confines of labour. In that training I made a case that labour leaders were also community leaders who could and should influence issues in their various communities. I was curious to know from the labour leaders why they thought that they couldn't be credible mediators between the communities and oil companies.

My question was specifically directed at labour leaders from the oil industry. It was at that training that I noticed that many leaders in the oil industry were from the oil-bearing communities. It was from this group that I partly developed my concept of critical constituency. Unfortunately, since that training I have not been able to follow up on these labour leaders and their possible role in transforming the conflicts in the ND. In conclusion there are a lot of local resources to be used for this peacekeeping force.

I have also had the rare privilege of working with several religious leaders in the ND. Rev. Fr. Andrew Ovienloba, the director of the Justice, Development and Peace Commission of the Catholic Archdiocese of Benin City in Nigeria has done a lot for peace in the ND. There are many other religious leaders who can be co-opted into this peacekeeping force. I recall with sadness the death of the Methodist priest who lost his life in the Umuechem crisis. He was among those who were already doing this work. All that is required is a proper coordination of their efforts.

The function of these groups will be as identified by Byrne and Keashly[17] - to enable parties in the conflict to get to know each other as individuals, for parties in the conflict to interact as equals, create a support base for institutional mechanisms for conflict resolution, to enable the parties to engage in cooperative tasks and to harness the available talents to develop positive attitudes of each other.

What will be the guiding principle of the work that these groups will do? My experience working in the ND suggests to me that they should operate from a transitional justice paradigm. Transitional justice incorporates three main values namely, truth, justice, and reconciliation[18]. And in this book I use the concept of transitional justice as a mechanism for groups or nations to evaluate their relations or in the case of states to evaluate their march to nationhood. In my conception, transitional justice incorporates such mechanisms as peace accords or agreements, writing of new constitutions, truth commissions, institutional reforms, elections, national conferences etc.

Using this paradigm the first assignment for the group will be to sign a symbolic national peace accord. This may sound grandiose or even empty, but there are precedents. There was no war during the transition in South Africa but they signed a national peace accord. Even in the ND, President Obasanjo in 2003 signed a peace accord with the Niger Delta Youth Movement (NDYM)[19]. Why the peace accord did not work is not within the purview of this book. But suffice to mention that because the process of signing the peace accord and its initiators lacked the core values of truth, justice and reconciliation, I am persuaded that this partly accounts for its failure. In the peace accord, which we are proposing, its main aim will be to end violence in the ND, if possible in the entire Nigerian nation.

After this the next move shall be to hold a conference of all groups in the ND. The main aim of this conference shall be for the world for the very first time to listen to the people of the ND. We pretend that we know what they want. We pretend to listen to them. But in actual fact this is not the case. This conference will lay out a broad framework for intervention in the conflicts in the ND. It will also be an avenue to gain the commitment of the people of the ND for the intervention. The main parties - namely the oil companies, government, NGOs and communities shall be represented at this conference. This gathering may also take the form of a truth commission where the people of the ND will be allowed to tell their stories. And the issues raised are addressed.

After this conference these same groups will embark on a fact-finding tour of communities in the ND. These visits will focus on oil spill sites, community development projects, erosion and pollution sites and all other such installations and sites which the people consider detrimental to their well-being. This is in line with the transitional justice principle which I mentioned earlier.

During these visits, the group will request and see all Memorandum of Understandings (MOUs) signed between the oil companies and governments. The main aim will be to know exactly what is outstanding with a view to addressing them later. It would be recalled that I made reference to a governor of a state who said that the conflicts in the ND was as a result of failed promises by the oil companies. A revisit of these MOUs will enable the group to put things into proper perspective coupled with their onsite visits.

The group will organize a symbolic gathering to remember all the people who have lost their lives in the cause of the conflicts in the ND. The importance of this is that to the people of the ND, all those who have sacrificed lives and limbs for the cause of the ND are martyrs and should be remembered as such. This move is anchored on the twin principles of truth and acknowledgement. Like Zehr[20] proposed in restorative justice, we must identify the harms done to the people of the ND and how these could be remedied.

Then this group will then embark on a process of reconciling communities, individuals, institutions and the general public. This will depend to a large extent on what the people want and their suggested approach to the process. It is important to note that these are suggestions that will focus on healing, rebuilding of relationships, after which all will now embark together on the project of peace-building. The steps and processes discussed above are aimed at ensuring that every one is included and participates in the task of conflict transformation and peace-building.

Notes and References

---

[1] See Sean Byrne, "Transformational Conflict Resolution and the Northern Ireland Conflict," International Journal of World Peace, Vol. XVIII, No.2, 2001, p. 3.

[2] Ibid; 4

[3] Ekiyor, Thelma. Women in Peacebuilding: An Account of the Niger Delta Women.From the Field, West African Network for Peacebuilding (WANEP) Issue No. 2, September, 2001. P.3

[4] Ibid: 5

[5] For a detailed discourse of "social cubism" see Sean Byrne and Neal Carter, "Social Cubism: Six Social Forces of Ethnoterritorial Politics in Northern Ireland and Quebec," Peace and Conflict Studies, Vol. 3, no. 2, 1996, pp.52-72.

[6]   Woodhouse, Tom and Oliver Ramsbotham (eds) Peacekeeping and Conflict Resolution. Portland, Oregon: Frank Cass Publishers, 2000, p.99.

[7] Ibid: p.97

[8] www.thisdayonline.com/news/20021121news05.html

[9] Okutimo was the head of the Internal Security force in the ND from 1994-1999.

[10] Vanguard Newspaper, August 6, 2004.

[11] Kriesberg, L. Constructive Conflicts, p.161.

[12] Frynas, J.G. Oil in Nigeria, p. 174.

[13]   Woodhouse, Tom and Oliver Ramsbotham (eds.) Peacekeeping and Conflict Resolution. Pp. 191-218.

[14] Ibid; 192

[15] John Paul Lederach and Paul Wehr, Mediating Conflict in Central America," Journal of Peace Research, Vol.28, No.1, 1991, pp.85-98.

[16] Op. cit. p.103.

[17] Ibid: 103-104.

[18] www.redress.org

[19] Interview with the Chairman and secretary of Abia branch of NDYM in 2003. The interview was recorded on video.

[20] Zehr, H. The Little Book of Restorative Justice. Pennsylvania: Good Books, 2000.

Chapter 7

A Model for Transforming Conflicts in the Niger Delta

In this chapter I shall share my experience of working in the ND. As mentioned in the introduction, I have carried out some interventions in the ND. The basis for this sharing is to show how some conflict situations have been transformed in the ND. The experiences, which I am about to describe, resulted from my work with the CSCR in Port Harcourt (PH). The CSCR started in PH basically as a human rights group but chose the name CSCR because according to its director "it is less threatening than any name with human rights in it"[1].

There are some basic assumptions about the work the CSCR does. First is that there is no need for the confrontations going on in the ND. Second is that both the oil companies, the government, and the communities have mismanaged relationships in the ND. Third is that the problem in the ND is partly because of the failure of the Nigerian State to live up to its obligations to its citizens. Fourth, is that NGOs in the ND have been less than thorough and altruistic in their various interventions. Finally and perhaps most importantly is that the oil companies need someone who will leverage them from inside. In other words the CSCR believed that many investors and staff of the oil companies do not agree with the conduct of oil companies in the ND. CSCR also assumed that the oil companies need reliable partners for them to do a good job in the ND.

A major gap, which I noticed in the work of CSCR, was that it had not perfected the strategies for interfacing with government agencies. Second, as an

NGO, its work was limited to Lederach's Track Two. This means that most efforts by CSCR will be concentrated at Track Three. Third, CSCR has not developed appropriate measures for combating the upsurge of violence in the ND. Fourth, the CSCR will have to develop confidence and trust-building measures between the parties. For instance most institutions, including government agencies, will open their doors to CSCR but not to the communities. CSCR cannot continue forever as a representative organization. It must learn to build the capacity of the local communities to engage both the government and the oil companies.

Finally, the CSCR must also begin to document its work with a view to sharing best practices. For instance, while I was researching this book, I could not lay my hands on documents to augment data from interviews and my personal experiences. On the whole CSCR's main handicap has been lack of funds.

Community Education and Institution-building (CEIB) Model

This model is inspired by three different approaches to change, which I have been exposed to. It will be better to say that these works gave form to my thinking and also refined my conceptualization of this model. The first is Lederach's elicitive training model which he discussed extensively in his book *Preparing for Peace*[2]. In this work, Lederach made the case that interventions must incorporate local resources. This means that an intervention is a learning process both for the intervener and the conflicting parties. In another work, Lederach referred to this as 'indigenous empowerment." According to him;

> "The principle of indigenous empowerment suggests that conflict transformation must actively envision, include, respect, and promote the human and cultural resources

from within a given setting. This involves a new set of lenses through which we do not primarily 'see' the setting and the people in it as the 'problem' and the outsider as the 'answer'. Rather, we understand the long term goal of transformation as validating and building on people and resources within the setting."[3]

Earlier on Curle has also alluded to this in his work at Osijek in Eastern Slovania. Curle observed that:

> "...since conflict resolution by outside bodies and individuals has so far proved ineffective [in the chaotic conditions of contemporary ethnic conflict - particularly, but not exclusively, in Somalia, Eastern Europe and the former USSR], it is essential to consider the peacemaking potential within the conflicting communities themselves." [4]

My coming in contact with these works emboldened me to further explore whatever was available in local resources. But more importantly Curle and Lederach's postulations appeal to me for two main reasons. First is that their thinking is a recognition of culture in conflict transformation processes. Second is that in a situation of acute lack of resources, one can draw from what is available locally. This second reason was a great motivating factor for my further exploration of this model.

The other work, which I drew from, was Kurt Lewin's *Action Research*. This methodology is used mainly in education curriculum planning. I found it very useful because of the in-built mechanism for self-reflection and the learning-as-you do-mode[5]. Lewin first wrote about action research in 1946 after the Second World War[6]. He used the phrase to describe research, which involves both the development of theory and advances in social change. Lewin defined action research (AR) as a form of research that begins with understanding and describing a situation in a particular context. This understanding and description is followed by an action plan. This process from "field of action" to "action plan" involves such principles as discussion, negotiation, exploration of opportunities, assessment of possibilities and an examination of constraints.

According to Lewin this stage is followed by the "action step" which is continuously monitored. During this action and monitoring stage, the researcher engages in learning, discussing, reflecting, understanding, rethinking and re-planning. This process is followed by an evaluation of the effect of the plan and action on the field of action. This evaluation in turn leads to a new action plan and the cycle begins again. In every day sense, the AR process is a five-step process. These steps could be translated to the conflict transformation field.

The first step will be to identify a conflict such as the one in the ND between the oil companies and the host communities. The second step is planning your intervention. The third stage is the intervention proper. The fourth stage will be the evaluation of the impact or effect of the intervention. The fifth stage is the sharing of the results and best practices of the intervention. This stage involves showing the relationship between the problems identified, the action taken, the information collected and analyzed and the evaluation of the action on the conflict. AR recommended itself to this model because of the in-built evaluation and reflective mechanisms. Second is that it involves participation by all those who are involved in the conflict. And it is an on-site kind of intervention.

I also drew a lot for this model from David Cooperider's *Appreciative Inquiry* (AI)[7]. AI recommends itself to me because of its implicit faith in the inherent goodness of individuals and organizations. AI arose as a critique of AR. In rethinking AR, Cooperider had argued that AR did not go "beyond merely secularized problem-solving frame"[8]. Cooperider in developing the conceptual framework of AI posited that our assumptions and the choices we make in interventions "largely create the world we later discover"[9]. AI therefore seeks to persuade us to be with, live with, and participate directly in our intervention process with a view "to inquire beyond superficial appearances to deeper levels of the life-generating essentials and potentials of social existence." This will help us to uncover the factors and forces that give life to the human spirit. In other words, AI argues that for every community, there is something good in them that have sustained it over time. In conflict situations, part of our mandate should be to uncover this life-giving force with a view to seeing how that can be used to build sustainable peace. An example is a couple that is in conflict. Instead of asking what is wrong, AI may try to find out what attracted the couple to each other in the first place. Second may be what has made them to live together for that length of time. The assumption here is that if you are looking for the good you find it, and if you looking for the bad, you will also find it.

AI like AR is also a step-by-step process. The four steps of AI are referred to as the 4Ds. The first D is to discover. This process involves discovering what is good about the community. In our context, what is good will be the local resources and culture that is available for peace-building. The second D is the dream stage. This is the stage where we envision the kind of community we want to have after conflict. The next stage is the design stage. It is during this stage that we plan how to make the dream come true. And finally we have the deliver or destiny stage. This is the implementation stage of the whole intervention. At the end of this stage, an evaluation is conducted and the process starts again.

I will now go on to describe the model which all of the above informed. My synthesis of AR and AI gave me what I shall refer to as Appreciative Action Research. A detailed analysis of this new concept shall be discussed in another book.

Meanwhile the name, which I gave to the model, is the Community Education and Institution-Building Model (CEIB), and a major ingredient of this model is what I will refer to as the Critical Constituency (CC). The CEIB model is designed for mainly community conflicts that are protracted. And as the name suggests it is used for conflict transformation. It is used for the prevention of outbreak of violence. It is very useful in high context cultures especially where the sense of community is gradually being eroded by mutual suspicion and disagreements. The model is supposed to unite the community, synchronize their views and create a united platform for engagement. It also emphasizes the traditional division of roles in communities.

CEIB is a process of bringing together members of a community in conflict with a view to educating them on the issues at stake and uniting them into one entity to engage a common issue. It is a very simple process. Members of a community are brought together at a symbolic venue that reminds them of common origin, ancestry, culture or history. And in this venue the issue confronted is laid bare and all aspects of it are examined and a common course of action is taken to engage the issue. For instance the gathering will examine the legal, economic, political and other angles of an issue. The main reason for the gathering will be to educate and elicit from the people a most acceptable response. The critical element here is a skilled facilitator.

That gathering becomes the front to address the issue. In this meeting as many people as possible will air their views, first by stating their understanding of the situation and second proffer possible solutions. And the facilitator will at all times find the areas of commonality of each viewpoint. But a consistent question throughout the meetings will be where do we go from here? There will also be a consistent sharing of responsibility and a feedback mechanism. The meeting should not be in a hurry to institute formal structures because this may lead to another possible conflict (the struggle for power). The emphasis at this stage should be to position the gathering as a service forum for the community. Leadership will emerge from those who are willing to take responsibility for carrying out agreed responsibilities.

This meeting which will run for sometime will be aimed at achieving two broad goals; (I) educating the community on the current issue with an eye on the future and (ii) formalising the organization as a permanent structure for the community to solve problems both for the immediate and the future. It also acts as a form of action research without really incurring the cost and burdens of a formal research. It educates both the intervener and the community at the same time. CEIB also builds the capacity of the community to takeover and manage its own affairs.

Origin, rationale, assumptions and context of CEIB

I began to think of CEIB when I became the sole administrator of my age set in 1984. I found out during my tenure that we learnt a lot from each other simply by being members of the age set. For instance, I remember vividly how we discussed the change of the colour of the Nigerian currency that year. I had just graduated from the university and my elder brother had just died. The age grade had come to condole with me. In my moody moments during my national service year I constantly reflected on that experience. That experience made a lasting impression on me. I became a very active member of my community development association. I was to be elected the secretary of one of the branches later. And in this position I harnessed the energy of the group for development. For instance my community was embroiled in a bitter conflict over who would succeed the late traditional ruler. The conflict had stagnated the community for a very long time, to the extent that whenever we attended the meeting of the association, it became the dominant issue of discourse. I saw how some people made a name for themselves by fuelling that conflict. Others became leaders simply by straddling the various parties to the conflict.

It was at this juncture that I decided to try out something. By this time every community development effort was either defeated or polarized along the lines of the conflict. I decided to change the mood of the branch meeting. I consulted widely and briefed people on my intention. Many were amenable to my idea, others were cynical, others sceptical while the rest were downright intransigent. It was a simple plan. We were going to give awards to schools in the community. This immediately changed the discourse in our meeting. When many argued against it, pleading lack of money, I told them that the gifts would be symbolic.

Anyway, we went ahead and gave out the awards - boxes of chalks and exercise books to the best students in mathematics and English language in the various classes. The symbolism of that simple gesture changed the dynamics of community development in my community. We became the branch to watch. The whole thing intrigued me even though I could still not put a framework on it.

The opportunity to put a framework on what I had experimented on came when I was appointed the Executive Secretary and Head of Conflict Resolution of my group. I began to mainstream community education in all my interventions. For instance we had so many cases of teenage pregnancy in the communities. In fact, in one month 22 cases were reported. Instead of tackling it I embarked on community education campaign of 15 different communities. I talked about the law on teenage pregnancy. Of course there was none. I explained the situation to the parents and community leaders. After that campaign I noticed a downward trend in the reported cases. Two years after the campaign there were several months that no single teenage pregnancy was reported. We did the same thing when there was cholera outbreak in one community and the police were arresting people and charging them for murder. I also embarked on community education and seized every opportunity to talk about their issues. They educated me on their understanding of the issues as well.

A few years later I was awarded the prestigious Ashoka Fellowship for my work in human rights. In my application for the fellowship I had described how I intended to harness the energy generated in conflict for community development. The Ashoka Fellowship ran for three consecutive, exciting but challenging years. It was during the fellowship period that I fine-tuned and put a framework on CEIB.

The main rationale for CEIB is that when communities engage in conflict, they do not often properly understand the issues at stake. This is especially so in a place like the ND where the issues have to do with the oil industry which is technical and complex. This has been the main reason why an entire community would contribute money and give to police as bribe when that same community could come together and constructively engage the police if they knew how. CEIB empowered the communities to take their destiny in their own hands. It also relieved the intervener of the responsibility of going back and forth, doing the same thing over and over again. For instance, an intervener may exit after intervention leaving the community to attend to such issues when they arise again. Another rationale was that since our work was not for profit and whatever we did depended on the availability of funds from the funding agencies; we tried to maximize the use of the available funds to institutionalize our work in the communities.

My main assumption in using CEIB was that conflict need not be divisive, and that when conflict is divisive, it is often deliberately constructed to be so by the elites in order to achieve a given end. I also assumed that conflict has a lot to do with lack of adequate information or what I call mal-information (that is the release of poor information on certain activities). I have come to believe that the most difficult decisions for people to make are those decisions where they do not have adequate information. Finally, I assumed that the prolonged period of military rule has divided our people and made to them lose their self-esteem. I also assumed that it was wrong for some people under the guise of representing a community during conflicts, decides to exploit them in the name of such representation. I also assumed that conflict intervention should be democratized especially when it involves a group.

I have found out over the years and through my use of CEIB that it works well in intragroup and intergroup conflicts. Though it could be used for interpersonal conflict it will require a lot of modification and adaptation. Another context for CEIB is for collectivist societies. CEIB works better in homogenous communities where people share meanings and symbolisms. I began to use it in the ND because most oil companies prefer to engage with community associations rather than individuals. CEIB in my opinion is democratic and encourages participation of all. It was used in the ND because most communities there have been torn apart by internal strife, mistrust and fractionalization. It was in an attempt to recreate the communities that was lost that we resorted to CEIB.

Let me also try to explain what I mean by institution-building. Among the people of Nigeria, having associations or groupings or what is now popularly known as civil society is part of the culture of the people. In some traditional Nigerian societies these groups rendered all kinds of services to the community. For instance certain age sets clear the path to the stream.

Others provided security, while others enforced decisions of the council of elders. With urbanization, some of the associations were established in the urban centres to provide support and sense of community for new arrivals in these areas. When the oil companies arrived, they decided that their best bet was to engage in what they referred to as community assistance programmes. This meant that they needed community Development Associations to work with. They even went as far as establishing Community Development Committees.

But like most things in the ND, these associations also became institutions of conflicts themselves. They became liabilities and were over-fractionalized. The power struggle in these associations was so debilitating to the communities that things literarily fell apart. Apart from the violence and bad blood, which the conflicts generated, they also damaged the fabric of the communities by pitching youths against elders, women against men and one section of the community against the other.

This situation made it almost impossible to carry out any meaningful intervention in the area. Even where the associations existed they were hijacked by those who had access to the infrastructure of violence. The leadership became a sit-tight one that was accountable to no one. There was no constitution, no election and therefore no correction. It was almost impossible to intervene in this kind of community. I remember going to one of the oil companies with some members of a community. On getting there, there were three other groups who had organized themselves to invade the meeting venue. The frustration I saw with the oil company staff was palpable and he asked, which group should I meet with? I said we were all representing the same community. After initial hesitation he agreed. And the meeting went well from hence. And if we had held the meeting without the other groups, it would have been very devastating to our reputation.

So by institution building I mean the process of organizing the members of a community into an association that will represent their interests and work for them. It also involves the process of writing a constitution, setting up an office, conducting elections and developing different files. It equally involves introducing the leaders of the association to the various agencies which they shall interface with later, building their capacity in the area of leadership as well as engaging democratic institutions, promoting human rights, community organization, gender equality, development, conflict resolution, peacebuilding and non-violence. In doing all these I was informed by the culture of the people. For instance the ND is a community full of different kinds of groups as we noted earlier.

Step by step process of CEIB

This is how I used CEIB: This could of course be adapted to the context, the organization's mandate and terms of reference. For CSCR where I used CEIB, part of the policy was not to intervene in a conflict except they were formally invited in writing. This may not apply to other organizations. The rationale for that very policy thrust is not within the purview of this project. Second,

CSCR was a not-for-profit organization, which will largely explain why CEIB recommended itself.

Before going into the details of the process, I also made extensive use of what I referred to as "critical constituency" in CEIB.

The story of how I came about this critical constituency (CC) is also a long and interesting one. I was conducting training for Catholic Youths all over Nigeria. In one of the sessions in the southeastern part of Nigeria I asked the participants the main objectives of the Catholic Youths Organization of Nigeria. They all chorused evangelism. I said fine, and I told them that if we studied riots and political violence in Nigeria, we could be sure that a large proportion of the participants were youths. They agreed with me. I told them that if went a step further and tried to find out what religious groups some of these youths belonged to that we might discover that some were Christians. They agreed. I said if we went a step further in our analysis, some might even be Catholics. They also agreed with me. Then I asked what they thought would happen if the CYON passed a resolution, even if symbolic, that no Catholic youth should be involved in violence like riots, political violence etc. I asked whether they thought that should be made an object of CYON. They agreed.

After that training I began to think about the number of Catholics youths in Nigeria and how dispersed the Catholic Church was in Nigeria. I came to the conclusion that this is a critical constituency in any community. My thinking was that for every community, there is a critical constituency. This model states that for every conflict there is also a critical constituency. This critical constituency is any group in a conflict situation that may cut across the divide of the conflicting groups or is an influential group or entity within the community. This may be a group of persons or idea or event or issue. This group must share something in common or the idea must be influential and symbolic enough to bind the people together and cut across the conflicting parties. For instance during the Eleme and Okrika conflict, both communities have Catholic adherents. You might even find officers of the CYON in both communities; my proposition is that we can use this tie to the same faith to build trust, confidence and even peace.

In the ND, the Catholic Church is very influential and the Catholic Youth Organization (CYON) is very prominent in the church. Therefore the CYON is a critical constituency both for analysis and intervention. In the CEIB model it is easy to use the CYON to reach out to other CYON members either within or outside the community. In fact, I was in a training recently where two officials of CYON were from the ND. Interestingly both were from different ethnic groups. If the two ethnic groups for instance were involved in a conflict, these two officials could become the CC we can use for peace-building.

An event that is a critical constituency could be something like the fishing festival. Because the ND is riverine, almost every community celebrates the fishing festival. CEIB could be incorporated into this festival. So the fishing festival is a critical constituency. The idea of being minority or a marginalized group can also be used as critical constituency. For instance if the people of the ND were made to realize that they are fighting for the same cause, then they will become united and engage the common oppressor collectively.

The challenge is for the intervener to have the presence of mind to identify these critical constituencies and use them strategically to educate, unite and build both the institutions and capacity of the community. It is also important to ensure that the critical constituency is one that cuts across the identity spectrum of the community and the conflicting parties.

Now let us look at the various steps of CEIB.

Step 1: The first step is that the community approaches the CSCR with a complaint. As mentioned earlier, CSCR does not solicit for cases and it does not take on cases except they are formally invited in writing.

Step 2: CSCR investigates the complaint. This is done through several channels. For instance, CSCR is a membership organization and has branches in most of the local communities in the ND. Moreover, CSCR networks with other organizations to check out information.

Step 3: Having certified the genuineness of the complaint, CSCR now approaches the community and explains in an open meeting how they work and how they intend to approach the issue. CSCR will also gain commitment from the entire community to engage in the process. When this is done, CSCR will confirm the internal resources the community has for peacebuilding and conflict transformation.

Step 4: CSCR will now approach the party who the complaint has been made against to open up a channel of communication. Here CSCR also confirms whether the party is aware of the conflict and what they have done about it. This could be another community, an oil company or their contractor or government agency.

Step 5: CSCR goes back to the community to inform them of the outcome of the meeting with the other party. It is from this point that CSCR will commence a series of institution-building activities. This involves several steps and activities. But they are all dependent on the nature of the conflict, the state of the community, the capacity and the level of fractionalization therein.

Step 6: This is basically training and capacity building and sharing of experiences, insights, stories and information. Here responsibilities shall be shared to members of the community. The key thing is that as many people as possible are involved in the process. But after the leaders are elected the level of attendance of other members of the community are likely to begin to decline.

Step 7: The institution is launched with as many people as possible invited to witness the ceremony. The other party in the conflict shall also be invited. This will be a good opportunity to meet with those that will work with them.

Step 8: The new executive with members of CSCR will pay a thank you visit to the organizations they invited to the launching, to thank them for attending or to find out why they did not honour the invitation. It is also during this meeting that they will secure an appointment to see the other party to discuss their issue.

Step 9: From here the community takes over the intervention process with CSCR 'eavesdropping' and lending a helping hand.

This is an annotated version of the model. The main issue to note is that the process is all-inclusive and participatory. Everyone shall be made aware of what is happening, and be part of what is going on. The process is also not a linear one. But it is important for the intervener to know at what point of CEIB he/she is. It is also important to note the exit points. At times CSCR provides logistical support at the initial stages of the intervention. CSCR also encourages the communities to always document their activities. They are also advised to have at least one- room office. We advice them against the use of an individual's place as office. We also try to discourage ad hoc meetings. We always propose that they have fixed (standing) days for their meetings.

There are many criticisms of this model. The first is that it wastes time. First, is the time to get the community together. Second, is the time to facilitate the meeting and achieve consensus. Third, there is always the possibility of the meeting becoming another talk shop with little or no action. These criticisms, though well founded, miss the point. There are basically three kinds of intervention. They are action, education and advocacy. CEIB incorporates all the three types and more. The extra is that it also creates a permanent structure for peace in every community. More importantly it is democratic. From the earlier chapters it is easy to see how the model fits the ND situation.

Another possible criticism is that critical constituencies have always existed and they may not have been very useful in peace-building. This is also true. But an issue that has come out from my studies is that critical constituencies in ND never saw themselves beyond the lenses of their traditional roles. Another possible criticism is that it fractionalizes the communities further. This may be true if the facilitator lacks the requisite skills. Secondly, the CC should not be used in isolation. It must be an integral part of CEIB.

Notes and References

---

[1] Interview with the Director of the Center for Social and Corporate Responsibility, Rev. Fr. Kevin O'Hara.

[2] Lederach, J.P. Preparing for Peace: Conflict Transformation Across Cultures. New York: Syracuse University Press, 1995.

[3] Lederach, "Conflict Transformation in Protracted Conflicts: The Case for a Comprehensive Framework, in Rupensinghe, K. (ed.) Conflict Transformation. Basingstoke: Macmillan, 1995, p. 212.

[4] Curle, Adam. 'New Challenges for Citizen Peacemaking, Medicine and War, Vol.10, No.20, 1994, p.96.

[5] Bray, John; Lee, Joyce; Smith, Linda L., and Yorks, Lyle (eds). Collaborative Inquiry in Practice: Action, Reflection and Making Meaning. Thousand Oaks, CA: Sage Publications, 2000.

[6] Lewin, Kurt. Action Research and Minority Problems, Journal of Social Issues, 2. 3446, 1946.

[7] Cooperider, David L. and Srivastva, Suresh. Appreciative Inquiry in Organizational Life. Research in Organizational Change and Development, Vol. 1. 1987, pp. 129-169.

[8] Ibid: p.1

[9] Ibid: p.1

Chapter 8

Reflections on the Niger Delta of Nigeria

In this last chapter I intend to lend my voice to certain concerns, which I have about the ND. These concerns are not new or peculiar to the ND situation rather it is one of the enduring dynamics of protracted conflict situations. I may not use many references here rather I will put out my observations, thoughts and insights especially as they affect the conflicts in the ND.

The first thing I will like to put upfront is that conflicts in the ND will be resolved and eventually transformed by ND indigenes themselves. I say this with a lot of caution. This is not to discount the idea of sharing, or that outsiders cannot help, rather I will like to emphasize the point that whatever intervention mechanism we put in place, if the people of the ND do not buy into it, it will not work and therefore will not be sustainable. This point is very important because as I was researching this project, I saw a lot of materials on the Internet requesting the international community to intervene. In fact one writer went as far as accusing the international community of coming to the rescue of Iraqis while ignoring the people of the ND. This is an ominous sign for conflicts in the ND. There are two possible frames to understand this. First is to describe it as a desperate call by a hopeless group. Another is to see it as a mischievous piece designed to attract the wrong kind of attention in the ND. Whatever happens the solution to the issues in the ND as we say it in Nigeria must be 'homegrown'.

Another observation, which I think we must put upfront is that, the ND issue is not only about money. It has very little to do with the revenue the people are getting from the federation account. Again I am not saying that they are getting enough or that they should not be given more. What I am saying is that simply giving them more money will be a naïve solution. This is because since 1960 (that is, since independence), there has been a gradual increase in the revenue due to the people of the ND. But instead of the conflicts subsiding in intensity and reducing in number, the opposite seems to be the case.

Let us now turn our attention to some of the more fundamental issues involved in the ND conflicts. Why is it that there is so much disunity among the people of the ND? I know that many have argued (albeit correctly) that some of the inter-communal conflicts have been fuelled by either government agencies or oil companies. I cannot dispute this. But since we now know that this situation could arise, what are the people of the ND doing to ensure that they may not be divided by those who do not have their interest at heart? To unify the people of the ND is a kind of intervention. The way to go about it is also a kind of model. That challenge is there, beckoning for those who have the presence of mind to work for peace.

Another side to this disunity and mistrust is the issue of proliferation of groups. I am also aware that the oil companies and the government have been accused of creating and financing these groups for their own selfish ends. But for how long shall this continue? Even if these groups have to exist independently due to the peculiar needs of their various communities, why can't they network and jointly articulate their issues? Critics will disagree with this observation arguing that they have been networking. But the evidence on the ground points to the contrary.

This issue is also very pronounced between the elites and the grassroots people. This division was one of the reasons some practitioners and scholars[1] with Marxist leanings have argued that the conflicts in the ND are embedded in a class struggle.

I have given my opinion on the conflicts in the ND. Suffice it to mention that this dichotomy between the elites and the grassroots has denied the ND the much-needed resources to mobilize and engage effectively. The dichotomy between elites and the grassroots is a recurring theme in history, but that does not mean that something cannot be done about it. It is therefore important for the conflicting groups to come together, especially in the interest of the coming generation.

I remember a very sad incident while I was conducting an intervention in a community in the ND. The community in question has complained about the number of scholarship awarded to them by the oil company. This complaint was based on a comparison with what a neighbouring community was getting. I had gone to the oil company to investigate why. Before this I had asked the community members the criteria used by the oil companies to determine the number of scholarships due to a community. Please note here that we are not talking about the actual scholarship but the application forms that qualify candidates to sit for the scholarship examination. On getting to the office of the person in charge of the scholarship, I asked him the criteria they use to determine the number of application forms due to a community. He told me that it was based on the number of barrels of crude oil produced in the community during the period. He did not expatiate on this. This did not bother me. I then asked him who the custodian of these figures was. He said the oil company. I now asked him how the communities could ever believe them considering the "litany of failed promises". At this juncture, he opened his drawer and brought out more application forms. Immediately a member of the delegation from the community grabbed these forms and pocketed them. This time I was visibly enraged and did not wait to leave the office before confronting the man who grabbed the forms. I said to him, "you are an engineer. You got your scholarship from this company. Now you work for another oil company. You have all the resources to train as many people as you want, why are you appropriating the applications forms meant for the widows, orphans and the poor in your community?" Of course he did not respond. He walked out with the forms, and as I write this, he has ceased to be part of our intervention in that community.

Another very troubling issue about the conflicts in the ND is that, everyone seems to know the solution, and in fact do sympathize with the people, but no one seems to want to implement those solutions. For instance I remember vividly when Olusegun Obasanjo's African Leadership Forum (ALF), in conjunction with Academic Associates Peaceworks did a lot of work in the ND. Since Obasanjo became president, he seems to regard the ND as an enemy to be conquered. This should not be surprising since as Burgess observes, it is one of the characteristics of protracted social conflicts. Even leaders from the ND find it difficult to commit fully to a change process that is in line with the vision of the kind of society envisaged by the people of the ND. More surprising was that it was under the late Bola Ige, as Minister for Justice and Attorney-General that the Federal Government went to court over the onshore/offshore oil dichotomy. Chief Bola Ige was quoted as saying that "all Nigerians are thieves, stealing the property of the Niger Delta. Nigerians have stolen the treasure of the Niger Delta people and if care is not taken, we will face the wrath of God because it is a sin to continue to plunder the resources of the people."[2] Bola Ige, in Nigeria's political landscape, belongs to that distinguished breed that is fast becoming extinct. How he allowed himself to be joined in this ignominy against the people of the ND we may never know. Later the government abrogated the dichotomy through legislation.

Closely related to the above observation of the inability of those in power to act in favour of the people of the ND, is the debate around why governments seem unable to control oil companies. Frynas[3] has identified corruption, the need to sustain flow of revenue to government coffers, and the vested interest of government officials in the oil industry as some of the reasons why governments seem unable to control oil companies. If this is the dilemma, what then is the solution to the capitulation of government in the face of the above reasons? Could this helplessness of government explain the hopelessness and subsequent desperation of the people of the ND?

The lesson from the above is that apart from treating the routine intergroup and interpersonal conflicts in the ND, there is the need for the people of the ND to network with people from other parts of Nigeria to engage in a systemic intervention for change. This is how some conferees captured this:

> "There is need for oil producing communities to form a common front and initiate mutually empowering alliance with social movements and other nationalities as a way of strengthening the capacity of individual communities and boosting the movement for democratization and auto-centric national development"[4].

Part of this process of forming "a common front" is development of a resource centre for information in the ND. I make this observation because when I was researching this project, it was difficult to pick and piece together information about the ND. For instance, it took more than three visits and almost six months for the people of Umuechem to produce copies of the bounced checks, which the government issued to them. There is no record of those who lost their lives in the Umuechem crisis. There is no record of those whose houses were burnt. It was even more difficult to ascertain from information on the ground whether the conflicts in the ND were subsiding or escalating. It would be interesting to have the record of all the youths arrested in oil-related "offences" in the ND.

The justification for this kind of information resource centre (which I call Niger Delta Resource Centre) is for us to be able to separate propaganda from public information service. This will also provide the avenue for the intellectualisation of the struggle of the ND against pollution and marginalization. Moreover, time has come to move the intervention in the ND a few notches up the pyramid of actors.

This common front may also involve the conceptualization of the ND as an economic unit. Right now the various governments in the ND, like governments in the other zones of Nigeria, are running around like a headless chicken. Part of the problem in the ND is that of unemployment. There are so many youths roaming around and ready and willing to be used. The various governments in the ND must conceptualize the area as an economic unit. For instance, the ND should have a thriving seaport, international airport, university and other investments that will not only symbolize the aspirations of the people of the area but provide employment for the teeming youths. I am aware that there is "a seaport and an international airport" and a host of universities in the area but these are 'meaningless' to the people of the ND.

But for the above to happen, violence must be minimized in the ND. I am not interested in who is responsible for what; the recriminations, accusations and counter-accusations have been going on for too long. The image of the ND as a violent area is not conducive to a project of redefining conflict resolution initiatives in the area. At times this image of violence is reinforced by people seeking for funds from the international community. The rhetoric of activists working in the area does not help matters either.

And to achieve this new image, there is a need to educate the people and distinguish between crime and social action. Ignorance has played a great role in endangering youths of the ND. The whole issues of bunkering, pipeline vandalisation, hostage-taking and all such crimes have been mistaken for instruments of the struggle. Now youths rob, steal, kill and destroy all in the name of fighting for their rights in the ND. Of course those that have been caught have paid a heavy price for it. Even innocent ones have been roped in. My first direct involvement with the ND was to bail out youths who were detained in Bonny. In the chieftaincy dispute which I referred to earlier in this book, ten youths were detained in prison in Port Harcourt awaiting trial for assault, one was in detention awaiting trial for murder over a chieftaincy dispute[5]. This case has now moved from the customary court to the Supreme Court.

Two other issues that arise out of this is the issue of who and how to engage for change. For instance since the present democratic experiment started, I have not noticed a concerted, consistent and coordinated effort by the people of the ND to engage democratic institutions. In democratic dispensations, legislative advocacy is a key tool for intervention. Unfortunately, this has not been used, at least to my knowledge. For instance it will be interesting to see which legislator(s) will oppose a revised revenue allocation formula in favour of the people of the ND. What has happened is that the conflicts in the ND have been drowned in the intrigues of partisan politics. This should not be so.

An issue that is related to the above is that of those we engage when we work in the ND. In Nigeria as a whole civil society organizations see themselves as Americans would put it "the good guys" while those in government or other organizations are seen as "the bad guys". This has led to the undue demonisation of all those in power. In short, civil society in Nigeria has developed some kind of anarchist ideology. I am not too sure of the exact source of this mentality but its impact on interventions has been very devastating. First, it has meant the engagement of just the communities, thereby shutting down communication with top officials of government and the oil companies. One interviewee for this project informed me that hobnobbing with these officials would compromise the intervener. I agree but that does not mean that we should exclude them.

I remember there was once an oil spill in a community and the oil company staff on ground wrote to the governor of the state accusing seven boys of being the ringleaders of those that sabotaged the oil pipeline. This accusation was made without investigation or proof. When we met with the top officials of the oil company and told them what their subordinates had done, they apologized and withdrew the letter[6]. The point I am making is that our interventions in the ND have remained at tracks two and three. But the fact is that those who facilitate the process of change are located in mainly track one. We need to develop the skills to engage those in track one. The Centre for Social and Corporate Responsibility (CSCR) has blazed the trail; we must share their best practices.

In this book I have made a case for the "de-politicization of conflicts in the ND". But I want to back up a bit and ask why is there no political party that articulates clearly the interests of the people of the ND? In Nigeria's peculiar political environment, people enhance their negotiating powers at the national level by leveraging their grassroots support. The Yoruba in Nigeria have never for once in Nigeria's political history been major players in a political party at the national level. They have consistently used their cohesion at the local level to get what they want at the federal level. The Igbos have also sometimes used this strategy. For instance in 1979, the Nigerian People's Party (NPP) won only three states in the federation but it was able to use that to go into an alliance with the National Party of Nigeria (NPN) at the national level. The people of the ND have never employed this strategy to put their issues on the political table of Nigeria.

I am not advocating for ethnic parties through the back door. A closer look at Nigeria's political history will reveal that people mostly speak either for their various regions or ethnic groups. At the just concluded Oputa Commission, all the representations were either Arewa, Oha na Eze, or Afenifere. Since this is the reality of the Nigerian political situation, the people of the ND must learn to speak with one voice. The betrayal of the resource control movement by some so-called leaders of the ND is still very fresh.

Another point, which I want to raise, is that of what I may call the "commercialization of the conflicts in the ND". Today there are so many people making money off the pain and agonies of people of the ND. The worst culprits are the oil companies, who, under the pretext of hiring conflict resolution consultants have brought in all kinds of people and used them to encourage capital flight. Locally we also have professional spokespersons who survive on the distress of the people of the ND. The international community has not helped matters by sponsoring all kinds of things in the ND. We need to ask this question again, why is it that in spite of all these investments, there is still no peace in the ND? I shall be interested in the near future to conduct a study on the amount of money that has been committed to training and conferences alone in the ND. And perhaps, study the impact of these on the situation in the ND.

The role of some ND indigenes who work in the oil companies is still an issue that puzzles me. For instance, I remember meeting an indigene of the ND who works in one of the oil companies. He was the Community Liaison Officer (CLO) for one of the communities. And in the meeting we held he was adamant defending the oil company where he worked. This may be natural. When I asked him whether he could re-state his position on the issue in his own community, he balked. This is just an example of the dismissive attitude of some ND indigenes that work in the oil companies. This is not to say that the job of changing oil company behaviour should be left to them. The argument is that they have the responsibility to educate their people on how to navigate their way around oil related issues in the oil companies.

If in my reflections I have been hard on the people of the ND, it is deliberate because I believe that the time has come for the people to rethink their strategies of engaging the government and the oil companies. The Nigerian government is not a neutral umpire in this whole scenario but as Dugan and Docherty have advocated, the people of the ND must address specific issues, then move to relational, and finally to systemic issues. The capacity to engage all the issues at the same time is not available - at least for now.

Notes and References

---

[1]Omoweh, David. The Role of Shell Petroleum Development Company and the State in the Underdevelopment of the Niger Delta of Nigeria. PhD dissertation submitted to the Obafemi Awolowo University, Ile-Ife, Nigeria.

[2] Quoted in Boiling Point, p.73.

[3] Frynas, J. G. Oil in Nigeria. Op. cit.

---

4 Committee for the Defence of Human Rights (CDHR). Boiling Point: The Crises in the Oil Producing Communities in Nigeria. CDHR: Lagos, 2000, p.263.

5 Interview with a youth from the area that the CSCR released from detention.

6 This intervention is still on-going as at the time of writing and because of ethical and security considerations the details shall remain anonymous for now. Any reader interested in the details should contact me.

Further Reading

Aboyade, Funke. "Resource Control an Answered Prayer" in Thisday Newspapers, www.thisdayonline.com/law/20021001law02.html

Achebe, C. The Trouble With Nigeria. London: Heinemann, 1983.

Agyeman, J. et al (Ed.) Just Sustainabilities: Development in an Unequal World. Cambridge, Massachusetts: The MIT Press, 2003.

Ake, Claude. The Feasibility of Democracy in Africa. Dakar, Senegal: CODESRIA, 2000.

Akinyemi, B. (Ed.). Readings on Federalism. New York: Third Press, 1980.

Amoo, S. G. "The Challenge of Ethnicity and Conflicts in Africa: The Need for a new Paradigm. UNDP, 2003.

Anderson, M. B. Do No Harm (www.cdainc.com)

Anifowose, Remi. Violence and Politics in Nigeria: The Tiv and Yoruba Experience. New York: Nok Publishers, 1982.

Ardrey, R. The Territorial Imperative; A Personal Inquiry into the Animal Origins of Property and Nations. London: Collins, 1967.

Augsburger, D.W. Conflict Mediation Across Cultures: Pathways and Pattern. Westminster/John Knox Press: Louisville, Kentucky, 1992.

Auty, R. M. Sustaining Development in the Mineral Economies: The Resource Curse. London: Routledge, 1993.

Awa, E.O. Issues in Federalism. Benin: Ethiope Publishing, 1976.

Azar, Edward. The Management of Protracted Social Conflict. Hampshire, England: Dartmouth, 1990.

Ballard, J.A. "Administrative Origins of Nigeria's Federalism", Journal of African Affairs, Vol. 70, No.281, 1971.

Bostock, W. Language Grief: A "Raw material" of Ethnic Conflict, in Nationalism and Ethnic Politics. Vol. 3, No. 4, 1994.

Browne, M. The Price of Oil: Corporate Responsibility and Human Rights Violations in Nigeria's Oil Producing communities. New York: Human Rights Watch, 1999.

Bray, John N., Lee, Joyce; Smith, Linda L.; and **Yorks, Lyle** (eds). Collaborative Inquiry in Practice: Action, Reflection and Making Meaning. Thousand Oaks, CA: Sage Publications, 2000.

Burton, J and Frank Dukes. Conflict: Readings in Management and Resolution. New York: St. Martin's Press, 1990.

Burton, J.W. (ed.) Conflict: Human Needs Theory. New York: St. Martin's Press, 1990.

Burton, J.W. Systems, States, Diplomacy and Rules. London: Cambridge, 1966.

Cable. S and Cable, C. Environmental Problems: Grassroots Solutions: The Politics of Grassroots Environmental Conflict. New York: St. Martins Press, 1995.

Chinweizu. The West and the Rest of Us. New York, Random House: 1978.

Clifford, Hugh: Address to the Nigerian Council on December 29, 1920. Document is available at National Library Lagos, National Archives Enugu and Ibadan.

Committee for the Defence of Human Rights (CDHR). Annual Report on the Human Rights situation in Nigeria. Committee for the Defence of Human Rights: Lagos, 2000.

Committee for the Defence of Human Rights (CDHR). Boiling Point: The Crises in the Oil Producing Communities in Nigeria. Lagos: Committee for the Defence of Human Rights, 2000.

Committee for the Defence Human Rights (CDHR). State Reconstruction in West Africa. Lagos: CDHR, 2001.

Constitutional Rights Project Report titled, "Land, Oil and Human Rights in Nigeria's Delta Region", 1999.

Constitution of the Federal Republic of Nigeria, 1999.

Cooperider, D. L. and Srivastva, Suresh. Appreciative Inquiry in Organizational Life. Research in Organizational Change and Development. Vol. 1, 1987, pp.129-169.

Curle, A. Tools for conflict transformation. London: Tavistock Press, 1971.

Daily Champion Newspapers, Saturday, September 14, 2002.

Davidson, Basil. The African Slave Trade. Boston: Cambridge University Press: 1961.

Delta Force, a documentary produced and broadcast on Channel 4, Ireland.

Deutsch, M and Peter Coleman. (Eds.) The Handbook of Conflict Resolution: Theory and Practice. San Francisco: Jossey-Bass, 2000.

Dibie, R. Understanding Public Policy in Nigeria: A 21st Century Approach. Lagos: Mbeyi & Associates, 2000.

Dike, K.O. Trade and Politics in the Niger Delta. London: Oxford University Press, 1956.

Docherty, J.S. Learning Lessons From Waco: When the parties bring their gods to the negotiation table. Syracuse, New York, 2001.

Douglas, O. and Doifie Ola, "Nigeria: Defending Nature, Protecting Human Dignity – Conflicts in the Niger Delta", in Searching for Peace in Africa, 1999, **www.euconflict.org.dev/ECCP**

Dudley, B.J. "A Coalition Theoretic Analysis of Nigerian Politics (1950-66), The African Review, Vol.2, No.4, 1974.

Ecumenical Council on Corporate Responsibility (ECCR). When the Pressure Drops. An Unpublished Report of a visit by ECCR (2001)

Environmental Rights Action (ERA/FoEN). The Emperor Has No Clothes. Port Harcourt: ERA, 2000.

Falola, Toyin. The History of Nigeria. Westport, Connecticut: Greenwood Press, 1999.

Fisher, R.J. The Social Psychology of Intergroup and International Conflict Resolution. New York: Springer-Verlaag Publishers, 1990.

Fortes, M. and E. E. Evans-Pritchard. African Political Systems. London: Oxford University Press, 1940.

Frynas, Jedrzej Georg. Oil in Nigeria, Conflict and Litigation between oil Companies and Village Communities. Hamburg: LIT, 2000.

Gary, Ian. Bottom of the Barrel. Baltimore, Maryland: Catholic Relief Services, 2003.

Gelb, Alan. Oil Windfalls: Blessing or Curse. New York: Oxford University Press, 1988.

Gurr, T.R. Why Men Rebel. Princeton, NJ: Princeton University Press, 1970.

Harbeson, J.W. and Donald Rothchild (Ed.) Africa in World Politics: The African State System in Flux. Boulder, Colorado: Westview Press, 2000

Harrison, L.E. and Samuel P. Huntington (Ed.). Culture Matters: How Values Shape Human Progress. New York: Basic Books, 2000.

Heinrich, Wolfgang. Building the Peace: experiences of Collaborative Peacebuilding in Somalia (1993-1996). Life and Peace Institute, 1997.

Human Rights Watch. "The Niger Delta: No Democratic Dividend", Human Rights Watch Country Report, Vol. 14, No.7 (A), October 2002.

Ikelegbe, A. "Civil Society, Oil and Conflict in the Niger Delta Region of Nigeria: Ramifications of Civil Society for a Regional Resource Control", in Journal of Modern African Studies, Vol. 39, No. 3, 2001.

Ikime, Obaro. The Fall of Nigeria: The British Conquest. London: Heinemann Publishers, 1982.

Imobighe, T.A. et al. Conflict and Instability in the Niger Delta: The Warri Case. Ibadan: Spectrum Books, 2002.

Isichei, Elizabeth. A History of the Igbo People. London: Macmillan Press, 1976.

Kieh, G.K. Jr. and Ida Rousseau Mukenge (Ed.). Zones of Conflict In Africa: Theories and Cases. Westport, CT: Praeger Publishers, 2002.

Kriesberg, L. Constructive Conflicts: From Escalation to Resolution. Maryland: Rowman and Littlefield Publishers, 2003.

Kruger, P. (Ed.). Ethnicity and Nationalism: Case Studies in Their Intrinsic tension and Political Dynamics. Marburg: Hitzeroth, 1993.

Kukah, M. H. Religion, Politics and Power in Northern Nigeria. Ibadan, Nigeria: Spectrum Books, 1994.

Lake, D.A. and Donald Rothchild, "Containing Fear: The Origin and Management of Ethnic Conflict", International Security, Vol.21, No.2, 1996.

Lederach, J.P. Preparing for Peace: Conflict Transformation Across Cultures. New York: Syracuse University Press, 1995.

Lederach, J.P. Building Peace: Sustainable Reconciliation in Divided Societies. Washington, DC: USIP Press, 1999.

Leith, Rian and Hussein Solomon, "On Ethnicity and Ethnic Conflict Management in Nigeria", African Journal of Conflict Resolution, No.1, 2001.

LeVine, R.A. "Anthropology and the Study of Conflict: An Introduction", The Journal of Conflict Resolution, Vol.5, No.1, March 1961.

Mack, R.W. and Richard C. Snyder, "The Analysis of Social Conflict – Toward an Overview and Synthesis", Conflict Resolution, Vol.1, No.2, June 1957.

Meek, C.K., "The Niger and the Classics: The History of a Name. Journal of African History, Vol. 1, No. 1, 1960.

Mitchell, C., "Beyond Resolution: What does conflict transformation actually transform?" Peace and Conflict Studies, Vol. 9, No. 1, May 2002.

Momoh, A. and Said Adejumobi (Ed.) The National Question in Nigeria: Comparative Perspectives. Burlington, Vermont: Ashgate Publishing, 2002.

Mwakikagile, G. Ethnic Politics in Kenya and Nigeria. Huntington, NY: Nova Science Publishers, 2001.

Neufeldt, Victoria (Ed). Webster's New World Dictionary of American English (Third College Edition). Prentice Hall, New York, 1994.

Nwabueze, B.O. A Constitutional History of Nigeria Harlow: Longman Publishers: 1982.

Nwagwu, M. "Meandering Pains of Resource Control" in The Channel Magazine, Vol.2, No.10, November 2001.

Okadigbo, Chuba. Power and Leadership in Nigeria. Enugu: Fourth Dimension Publishers, 1987.

Okonta, Ike and Oronto Douglas. Where Vultures Feast: Shell, Human Rights and Oil in the Niger Delta. San Francisco, CA: Sierra Club Books, 2001

Osaghae, E.E. Crippled Giant: Nigeria Since Independence. Bloomington, Indiana: Indiana University Press, 1998.

Osaghae, E. E. "Ethnic Minorities and Federalism in Nigeria". African Affairs (1991), 90, 237-258

Osaghae, E. E. "The Ogoni Uprising: Oil Politics, Minority Agitation and the Future of the Nigerian State." African Affairs, 94, 1995. pp. 325-344.

Osaghae, E. E. "Managing Multiple Minority Problems in a Divided Society: Nigerian Experience." Journal of Modern African Studies 36 (1), pp. 1-24, 1998

Perham, Margery. Lugard: The Years of Authority (1898-1945). London: Commonwealth Publishers, 1960.

Pryor, B. Ethnographic Study in Environmental conflict Resolution: The Role of Environmental NGOs in the Florida Everglades Restoration. (An unpublished PhD dissertation), 2003, Submitted to the Department of Conflict Analysis and Resolution, Nova Southeastern University, Florida

Ritzer, G. and Douglas J. Goodman. Sociological Theory (6th Edition). New York: McGraw Hill, 2004.

Rodney, Walter. How Europe Underdeveloped Africa. London, Bogle-L'Ouverture Publications: 1981.

Ross, Michael. How Does Natural Resource Wealth Influence Civil War? Los Angeles, CA: UCLA, 2001.

Rothman, J. Resolving Identity-Based Conflict in Nations, Organizations and Communities. Jossey-Bass: San Francisco, 1997.

Saro Wiwa, Kenule. On a Darkling Plain. London: Epsom, 1989.

Schulz, A. The Phenomenology of the Social World. Northwestern University Press, Evanston, Illinois, 1967.

Sunday Champion Newpaper, November 25, 2001.

Tebekaemi, Tony (Ed.). Major Isaac Jasper Adaka Boro: The Twelve-day Revolution. Benin City: Idodo Umeh Publishers, 1982.

Thisday, Friday January 5, 2001.

Thisday, Friday February 2nd, 2001.

Thisday Newspaper, Friday, February 2nd, 2001

Thisday on Saturday, February 24, 2001

Thisday, November 28, 2001,

Thisday on Sunday, January 18, 2004.

Thisday Newspaper, July 19, 2004

Thomson, **A.** An Introduction to African Politics. London: Routledge, 2000.

Torulagha, P.S. "The Niger Delta, Oil and Western Strategic Interests: The Need for an Understanding" (www.unitedijawstates.com/niger_delta.html)

United Nations Development Programme (UNDP) Report, "Promoting Conflict Prevention and Conflict Resolution Through Effective Governance", 2003.

Udo, R.K. Land Use Policy and Land Ownership in Nigeria. Lagos: Ebieakwa Ventures Ltd, 1990.

Uwazie, E.E. et al (Ed.). Inter-Ethnic and Religious Conflict Resolution in Nigeria. Lanham, MD: Lexington Books, 1999.

Welch, C.E., Jr. The Ogoni and Self-determination: Increasing Violence in Nigeria. The Journal of Modern African Studies, 33, 4 (1995).

Williams, D.C., "Measuring the Impact of Land Reform in Nigeria", The Journal of Modern African Studies, 30, 4 (1992).

Woodhouse, Toma and Oliver Ramsbotham (eds.). Peacekeeping and Conflict Resolution Portland, Oregon: Frank Cass Publishers, 2000.

www.thisdayonline.com/news/20031224news05.html

Zehr, Howard. The Little Book of Restorative Justice. Pennsylvania: Good Books, 2000.

Index

P

O

Q

www.ingramcontent.com/pod-product-compliance
Lightning Source LLC
Chambersburg PA
CBHW020608270326
41927CB00005B/230